Urbanization and
Cancer Mortality

Monographs in Epidemiology and Biostatistics
edited by Abraham M. Lilienfeld

Monographs in Epidemiology and Biostatistics
Volume 4

Urbanization and Cancer Mortality

The United States Experience, 1950–1975

Michael R. Greenberg

PROFESSOR OF ENVIRONMENTAL PLANNING AND GEOGRAPHY
RUTGERS UNIVERSITY

New York   Oxford
OXFORD UNIVERSITY PRESS
1983

Copyright © 1983 by Oxford University Press, Inc.

Library of Congress Cataloging in Publication Data

Greenberg, Michael R.
    Urbanization and cancer mortality.

    (Monographs in epidemiology and biostatistics ; v. 4)
    Includes bibliographies and index.
    1. Cancer—United States—Mortality.   2. Cancer—
Mortality.   I. Title.   II. Series.   [DNLM: 1. Neoplasms—
Mortality.   2. Urbanization.   3. Neoplasms—Mortality—
United States.   4. Urbanization—United States.   QZ 200
G798u]
RC276.G73 1983   614.5'9994'00973   82-7942

ISBN 0-19-503173-3

Printing (last digit): 9 8 7 6 5 4 3 2 1

Printed in the United States of America

To Alexandra Greenberg

Preface

Urbanization and cancer have both become more important in twentieth-century life in the United States. In 1900, only about 40 percent of the American population lived in urban areas. Cancer was then only the eighth leading cause of death in the United States, responsible for less than 4 percent of deaths. Pneumonia, influenza, and tuberculosis each caused about three times as many deaths as cancer.

Seventy-five years later, the United States was an urban nation with much higher cancer death rates. By 1975, more than 70 percent of the population resided in urban areas, many in metropolises along the east and west coasts and the Great Lakes. Cancer had become the second leading cause of death, accounting for more than 350,000 deaths each year and almost 20 percent of all deaths. Although heart disease was responsible for more deaths than cancer in 1975, among the major diseases, only cancer has not shown an obvious long-term decline during the twentieth century.

Research and arguments about the relationship between urbanization and cancer have focused on pollution. Notwithstanding the importance of the pollution issue, it was and is a mistake to have so narrowly focused the urbanization and cancer debate because the pollution and cancer link is so difficult to prove and because the focus of the debate on pollution has drawn attention away from differences between urban and rural areas in cigarette smoking, alcohol use, and occupational exposures and probably nutritional habits, stress, and medical practices. Recasting the urbanization and cancer link in historical and cultural contexts allows personal habits and occupation as well as pollution to be included in the urban factor. In turn, broadening the urban factor provides a much firmer foundation for explaining the most pronounced trend in the geography of cancer mortality in the United States: increasing homogeneity.

The purposes of this book are to show that upon controlling for race, sex, and age the United States has been moving toward a spatial convergence of cancer mortality and that the most probable explanation of this tend is that the United States has become an urban-oriented culture characterized by increasingly similar smoking, drinking, and nutritional habits, and occupational and other risk factors in villages, small towns, and suburbs as well as cities. The book is organized as follows. The first chapter reviews the urban factor literature focusing on those types of cancer that have been most consistently identified to manifest higher rates in urban than in rural areas and on arguments for an urban pollution factor. Chapter 2 presents cancer mortality data for 1950–75 and methods for changing these data into indices for determining the changing relationship between urbanization and cancer mortality. A cancer mortality base line for the United States and other nations is provided in Chapter 3. Age-adjusted rates for white male, nonwhite male, white female, and nonwhite female Americans are compared for 35 types of cancer for the time periods 1950–54, 1955–59, 1960–64, 1965–69, and 1970–75. Chapters 4 through 6 set forth the changing geographical distribution of cancer mortality in the United States. Chapter 4 compares cancer mortality rates and general indicators of cancer risk in urban and rural areas of the United States and compares cancer mortality rates and urbanization in other nations. Child, teen, and total population cancer mortality rates among the most populous, metropolitan areas of the United States are compared in Chapter 5. Chapter 6 presents the decreasing differences in cancer mortality profiles between the cities and their suburbs. The last chapter is a summary and discussion.

A few words of explanation are offered to specialists who may be disturbed by where some information has been placed and by the omission of some information. Those of us who enjoy having supporting tables on the same page as the text are apparently a small minority. Accordingly, 50 tables have been shifted to appendices as well as the explanation of some methods and a comparison of mortality and incidence rates. Finally, I apologize to my medical geography, medical sociology, and environmental science colleagues for not labeling the origins of all the research presented in the book. This omission was made in order to spare those many readers who are not interested in the disciplinary origin of research.

Highland Park, New Jersey   M. G.
August 1982

Acknowledgments

The thesis presented in this book began with four research projects for the Office of Cancer and Toxic Substances of the New Jersey Department of Environmental Protection. I owe a debt to the State of New Jersey for funding investigations into the changing geography of cancer mortality in the northeast United States. I would like to thank Drs. Glenn Paulson, Peter Preuss, Sidney Gray, Judith Louis, Thomas Burke, and Paul White of the New Jersey Department of Environmental Protection for discussing the implications of the geography of cancer in the New Jersey region with me during our four year association.

Bill McKay of the National Cancer Institute and John Caruana of Rutgers University were responsible for much of the data processing. Without their computing skills, it would not have been possible to sort millions of cancer deaths into appropriate regional configurations.

During the last six years, I have taught a class at Rutgers University in applying statistical methods to toxic substance and cancer data. Two parts of this book were products of these classes. Specifically, Kelly K'Meyer worked on the discriminant analysis that appears in Chapter 5 and Joyce Batchelor, Joseph Benz, and Dona Schneider worked on the childhood and teenage cancer mortality analysis in the same chapter.

Jeffrey House and Carol Miller of Oxford University Press are acknowledged for their extremely careful reading of the manuscript, and I would also like to thank Drs. Lynne Harrison, George Sternlieb, and Briavel Holcomb for their comments about specific sections of the book.

Special thanks go to Sharon Waldman, Vera Lee, and Kathy Bignell for typing portions of the manuscript.

Contents

Urbanization and
Cancer Mortality

1 The Urban Factor in Cancer: Previous Reports

An association between urbanization and mortality has been found for . . . cancer of the respiratory and digestive systems, and many other diseases, but the reasons are not fully understood. This urban-rural difference results partly from self-selection and errors in reporting place of residence—many people who become seriously ill move to cities where they can obtain better medical and hospital care, find less demanding occupations, and lead more sedentary lives.

Even when allowance is made for these factors, there remain higher mortality rates for many causes of death, including cancer. This excess risk may be related to life-style (urban dwellers use more tobacco and alcohol), to occupation (working with industrial pollutants), to the environment of the cities themselves (air and water pollution, etc.), to other factors not yet identified.

DAVID LEVIN et al. 1974

The insertion of a variable representing either urbanization or population density into epidemiological mortality models has become almost as standard a procedure as including an income variable in econometric models, but the usage is quite different. The income effect has been thoroughly investigated in economic theory and adjustment for its influence is well understood. On the other hand the urban effect in mortality models has been observed but not well analyzed, and its influence attributed to a number of associated characteristics ranging from measureable and specific to unmeasurable and vague, such as "tension, stress, and unhealthy personal habits." The inclusion of an urban variable in order to adjust for all its associated factors, both known and unknown, essentially begs the question. What is needed is to determine which components of the urban factor, and in what combinations and relative strengths, best explain the . . . mortality patterns for different types of cancer.

DOROTHY GAITES WELLINGTON et al. 1979

3

It is conventional wisdom that there is an "urban factor" in cancer, and there is good reason for this, namely, consistent evidence of higher rates in urban than in rural areas. This finding, however, remains poorly understood. Many interpret it as reflecting widespread air and water pollution in urban areas, although the overwhelming majority of scientists attribute less than 5 percent of cancers to pollution (Office of Technology Assessment, 1981). The view presented in this book is that the higher rates of some types of cancer in urban than in rural areas are the result of historical and geographical processes that led to the concentration of such important risk factors in cities as industrialization, high rates of smoking and drinking, a Western-world diet, Western medical practices, worker exposure to dangerous substances, and environmental degradation.

If this is correct, then, since many of the characteristics of Western urban living are now widespread outside the cities, cancer rates should have become independent of cities. Surburban and rural areas should increasingly show a pattern of disease formerly identified only with cities.

This chapter reviews the large and complicated literature on the relation between urbanization and cancer, organized by major cancer types: respiratory, digestive, female breast, and urinary. It deals with some studies in detail and many others more briefly.

Three basic observations can be drawn from the literature. First, although many types of cancer have been reported to manifest an urban excess, only respiratory (especially lung), digestive (especially intestine), female breast, and urinary (especially bladder) cancer have been documented with sufficient frequency and consistency to warrant the inference that higher rates exist in urban areas than in rural areas.

Second, there is no overwhelming evidence either for or against an urban pollution factor in cancer. Even studies with noteworthy findings have methodological weaknesses. Many of these studies have been criticized for the following reasons: (1) the geographical regions being analyzed are extremely large, and therefore, there is a potential of the "ecological fallacy," extrapolating results from aggregated population groups to individual cases (see Chapter 2 for a more detailed discussion); (2) important intervening variables, such as the rate of cigarette smoking, are usually not included in the analysis; (3) the indicators of the risk factor are almost always too general, such as the measurement of total suspended particulates, rather than the amount of a specific carcinogen such as benzo(a)pyrene in the polluted air; (4) the long latency period of cancer (typically two to four decades after initial exposure) is not taken into account; and (5) too many unexpected and therefore probably spurious relationships are revealed by the studies, but are not addressed by the authors. In addition, toxicological studies have shown that samples of air, water, and soil and of fish and animals taken in the outdoor, workplace, and home environments in urban areas contain carcinogens. There have been few thorough environmental surveys, however, and the conclusions drawn from studies involving a few samples are considered to be inadequate for firm inferences to be made.

Third, the literature is marked by affirmations and denials. An excess of cancer, or of carcinogens in the environment is reported, and the finding is confirmed by other scientists. Soon afterward, however, alternative explanations of the phenomenon are offered and/or the data and methods used in the original studies are found wanting. In addition, it is not uncommon for researchers to draw different and sometimes contradictory conclusions from the same data, or from different data for the same region.

RESPIRATORY SYSTEM

The issue of increased rates of respiratory cancers, and especially lung cancer, in urban areas and their relationships to air pollution has been hotly debated. In 1932 Tobey claimed that

> Endeavors have . . . been made to show that smoke may have some influence on cancer, but there is actually no reliable data to incriminate smoky atmospheres, undesirable as they may be from the standpoint of general hygiene.

Four decades later, a National Academy of Sciences Conference on the Health Effects of Air Pollution (1973) concluded that

> There are substances present in polluted atmospheres which have been demonstrated in experiments on animals to be mutagenic or carcinogenic. The concentrations required in such experiments have been very much higher than those which might be encountered in polluted atmospheres. There is an urban-rural gradient in human lung cancer not attributable to cigarette smoking. The atmospheric pollutants here considered may be among the factors contributing to the higher rate in urban areas but definitive evidence is lacking.

Following the torturous route of the literature from the 1930's to the 1980's is the best way to understand this long-standing controversy. The first important paper is by Stocks and Campbell (1955), who found urban cancer mortality rates in mid-1950 England to be twice as high as rural cancer mortality rates. Lung cancer rates were much higher in densely settled portions of urban areas in England and Wales than in more sparsely settled, rural areas. Stocks and Campbell attributed 50 percent of the lung cancer rate in Liverpool to smoking and 40 percent to air pollution.

Similar findings were soon reported elsewhere. In 1956, Haenszel (Hagstrom et al., 1967) found significantly higher rates for lung and larynx cancer in the urban areas of Iowa than in the rural areas. In a study of about 200,000 white American males aged 50 to 69, Hammond and Horn (1958) found that the age-standardized lung cancer rate for those residing in cities with a population of over 50,000 was 52 per 100,000, whereas the rural rate was 39. Levin (1960) found higher rates of cancer of the respiratory system (as well as of the intestine, rectum, and esophagus) in urban compared with rural areas of New York State. Haenszel, Loveland, and Sirken (1962) and Haenszel and Taeuber (1964), focusing on smoking habits, found that males who smoked more than one pack of cigarettes a day had a lung cancer mortality rate 31.4 times higher than males who had never smoked. They were able to define differences in risk according to number of years of smoking, numbers of cigarettes smoked, tar and nicotine content of cigarettes, and inhalation practices. In addition, they found that lung cancer mortality rates in urban areas were 43 percent higher than in rural areas. Furthermore, the longer the residence in urban areas, the higher the lung cancer rate, which implied that air pollution might be a factor. Other significant associations between air pollution and cancer death rates were reported during this period by Manos and Fisher (1959), Prindle (1959), and Schiffman and Landau (1961).

The first mortality reports on migrant populations also appeared during the mid 1950's and early 1960's. Dean (1961) found that white migrants from Britain to South Africa had a cancer mortality rate 44 percent higher than white men born in South Africa. Eastcott (1961) reported the lung cancer rate among British migrants to New Zealand to be about twice as high as that of native New Zealanders. In both studies, the data were controlled for age and smoking habits.

Initial evidence about specific carcinogens in the air and their link to cancer was also collected during this period. Stocks (1960) found that lung cancer and bronchitis were significantly related to the presence of the carcinogen 3,4-benzo(a)pyrene in the air in parts of Britain. In response to this and other work, a symposium was held in 1961 on the determination and measurement of chemical carcinogens in the air (Sawicki and Westphal, 1961).

It was not long before hypotheses that urban/rural differences in lung cancer rates could be explained by factors other than air pollution were advanced. Wynder and Hammond (1962) suggested the following alternatives: (1) urban/rural differences in smoking patterns; (2) more accurate reporting of illnesses in urban than in rural areas; and (3) occupational differences. They noted that urban/rural differences were small compared to smoker/non-smoker differences. As indicated above, Hammond and Horn (1958) found that, after standardizing for age and smoking habits, male lung cancer mortality rates were 25 percent lower in rural areas than in large cities. They concluded, however, that "whatever the urban factor may be, its effect . . . is small as compared with the effect of cigarettes." Doll (1963) and Doll and Hill (1964) also argued that the role of atmospheric pollution and occupational exposures in producing lung cancer is small compared to the role of smoking. They estimated that lung cancer in Britain would be reduced to 11 percent of the existing rate if people did not smoke. In a later paper, Doll (1970) referred to the fact that lung cancer was much higher in Britain than in other English-speaking countries as the "British factor," rather than the air pollution factor. Doll's arguments for the clearly dominant role of smoking in carcinogenesis continue to be strongly stated (Doll and Peto, 1981) and have been echoed by others (e.g., Higgins, 1976; Waller, 1972; Marmor, 1978).

There were sharp responses to these arguments. Drawing from over 1,000 papers and books, Hueper (1966) argued that industrial processes and polluted air, water, and land were important sources of carcinogens. The vehemence of his arguments is evident in the two following quotations:

> It should be obvious that any wide acceptance of such scientifically unsound and socially irresponsible claims concerning the principal role of cigarette smoking in the causation of cancers, especially respiratory cancers, would paralyze not only a legitimate and urgently needed pursuit into the various environmental factors inducing such cancers, particularly the many industry-related pollutants of the urban air, but has provided already effective legal arguments before civil courts and compensation boards for denying justified

claims for compensation of occupational respiratory cancers to the victims of such hazards as well as to their widows and orphans.

The recent allegation ... that lung cancers in chromate workers were found only in individuals who smoked cigarettes thus appears to be an expedient subterfuge and skillful device helpful to industrial management for escaping their legal and moral obligations toward their employees and human society. Such fanciful contentions have obviously no scientific basis in fact and belong to the pseudo-scientific propaganda emanating during the last decade to an increasing degree from some scientific guardians of commercial interests.

Like Hueper, many authors have reported an association between occupation and lung cancer (e.g., Newman et al., 1976; Fox et al., 1974; Mabuchi et al., 1979; Blot et al., 1979; Shigematsu and Yamasaki, 1978). The most alarming case was made by a group from the National Cancer Institute and the National Institute of Environmental Health Sciences (1978), who argued that worker contact with asbestos, arsenic, coal tar pitch volatiles and coke oven emissions, chromium, iron oxide, nickel, and petroleum distillates could account for 20 percent or more of total cancer incidence and a much higher percentage of lung cancer incidence. Doll and Peto (1981) called this a paper that "should not be regarded as a serious contribution to scientific thought and should not be cited or used as if it were," and attacked the methods used to make the estimate; they attributed 4 percent of cancer in the United States to occupational exposure.

Thus, during the 1950's and early 1960's, the first epidemiological and toxicological studies showing an urban excess of lung cancer that might be related to air pollution were published. They were countered by arguments that cigarette smoking clearly was the most important environmental factor contributing to respiratory cancer and that, together with occupational exposures and data and methodological shortcomings, smoking explained the urban excess.

During the late 1960's and 1970's, several case studies provided evidence both for and against a relationship between air pollution and lung cancer. The first two studies discussed here are from California and show how different approaches may yield different results for the same area. Buell, Dunn, and Breslow (1967) reported negative findings on air pollution as a cause of respiratory cancer in a study of lung cancer mortality rates among California legionnaires in Los Angeles, San Diego, San Francisco, and other California communities. Controlling for smoking patterns and length of stay in their present community of residence, the investigators unexpectedly found that lung cancer rates in Los Angeles were lower than those in San Diego and San Francisco. However, they suggested that their study period of 1959–62 was perhaps too short for photochemical air pollutants to have an effect. Buell and Dunn (1967) stated that although two studies have "failed to find an urban-rural difference in nonsmokers . . . the more numerous positive results cannot be explained by chance." A few years later, such positive results came from another Los Angeles study. Menck et al. (1974) grouped the 26 health districts of Los Angeles into 13 areas with homogeneous air pollution profiles and controlled for socioeconomic status. They reasoned that cigarette smoking and airborne polynuclear aromatic hydrocarbons from petroleum and chemical industries could explain the high rates of lung cancer they found in several Los Angeles health districts.

The pattern of the two California studies is repeated in many others from the mid 1960's into the 1970's. Hagstrom, Sprague, and Landau (1967) did not find a strong relationship between air quality and lung cancer in the Nashville, Tennessee region, but called for additional studies because of data shortcomings. Greenburg et al. (1967) suggested that excess cancer deaths in Staten Island, New York were due to air pollution, but did not control adequately for smoking, socioeconomic status, and other variables. Lave and Seskin (1970), reworking data from England and Wales, estimated from regression equations that lung cancer would decline between 11 and 44 percent if all boroughs had the air quality of the borough with the best air. They produced equally striking equations using American data. When occupation, climate, and home heating were added as variables to the equations (Lave and Seskin, 1971, 1972), the effect of air quality on mortality rates decreased. However, Lave and Seskin argued that since occupation and home heating are sources of air pollution and since the climate determines heating and cooling requirements, the overall relationship between air quality and health is strengthened. Weiss (1978) reported a positive association between air pollution sources and male lung cancer mortality rates in Philadelphia, but cautioned against cause-and-effect conclusions because of the absence of smoking and occupational data. Associations between air quality and cancer have been reported in similar studies by Koshal and Koshal (1973), Page (1973), and MacDonald and Schwing (1973). All these studies, however, can be criticized for one or more of the five different methodological problems set forth earlier in the chapter.

To strengthen the case for air pollution as a cause of lung cancer, repeated evidence of high concentrations of carcinogens in the air is required, as are epidemiological studies linking these substances in the air with cancer. Such studies have been done since the late 1960's. Hydrocarbons, and to a lesser extent asbestos, have been studied most extensively. As noted, benzo(a)pyrene, a polycyclic aromatic hydrocarbon, was the early focus. Indeed, studies of the relationship between air pollution and cancer have been based to such a degree on benzo(a)pyrene data that Lawther (1965) suggested that attention be shifted to other substances.

A strong circumstantial case can be made for polycyclic aromatic hydrocarbons as being the major carcinogen. They are produced by industry (Gerrard, 1970), by motor vehicles (National Academy of Sciences, 1972), and by heat and power generation and refuse burning (Council of Environmental Quality, 1976). Benzo(a)pyrene extracted from city air proved to be a strong mutagen to mice (Epstein et al., 1966) and polycyclic aromatic hydrocarbons produced skin cancers in mice (Falk et al., 1964).

Furthermore, the National Academy of Sciences (1972) suggested that polycyclic aromatic hydrocarbons may interact with ozone and other photochemical oxidants, increasing the impact of each. Although more than 80 percent of the polycyclic aromatic hydrocarbons in the air come from heat and power generation and refuse burning (Council of Environmental Quality, 1976), in urban areas, emissions from motor vehicles may be their most important source. Pike et al. (1975) claim, for example, that motor vehicles produce virtually all the benzo(a)pyrene in the air of downtown Los Angeles.

Although polycyclic aromatic hydrocarbons are commonly found in the air and are carcinogenic (Grover, 1973; Rigdon and Neal, 1971), their role as the dominant carcinogen in polluted air remains questionable because of periodic negative results and because of the presence of other carcinogenic air pollutants. For example, Spangler and DeNevers (1975) examined the air and soil in Charleston, South Carolina for the presence of benzo(a)pyrene and trace metals because Charleston and other Southeastern coast cities have high lung cancer mortality rates. They found concentrations of benzo(a)pyrene in the air and soil to be lower than the national average. The relationships are made more complicated because laboratory studies (Pike et al., 1975) indicate that the enzyme aryl hydrocarbon hydroxylase (AHH) is involved in converting benzo(a)pyrene to a carcinogen. However, AHH activity is thought to be very different in women than in men, and Pike argues that as a result the "relation of air pollution to lung cancer in females is very unclear." The picture is also complicated by the fact that other hydrocarbon and inorganic carcinogens (such as asbestos) have been found in the air and must also be considered (Gordon et al., 1973; Asahima et al., 1972). Thus, although living in a major urban area may be the equivalent in exposure to benzo(a)pyrene of smoking between two and six cigarettes a day and although polycyclic aromatic hydrocarbons may be the most obvious carcinogenic substance in the air, they may not be the most important. Their role is not yet clear.

Cooper (1967) noted that "asbestos would not be in the headlines if asbestosis, affecting a few hundred miners, millers, and insulating workers, were the only problem." Asbestos has been used in more than 3,000 products (National Academy of Sciences, 1971; National Research Council, 1971; Hallenbeck et al., 1977), including brake linings, and road-surfacing materials, and it is mixed in floor tiles and in textured paints used in many public buildings (schools, hotels, and offices) and in homes.

The widespread use of asbestos in consumer products is matched by its widespread presence in human populations (Cooper, 1967; Selikoff, 1968; Hagerstrand and Seifert, 1973). Cooper (1967) noted in seven clinical studies made between 1963 and 1966 that a range of 26 to 58 percent of the general population in such places as New York City, Cape Town (South Africa), Pittsburgh, Montreal, and Miami was found to have "mineral-fibre bodies" in their lungs; some of these bodies were undoubtedly asbestos. Selikoff, a leading expert on the health implications of asbestos, reported in 1968 that 702 of 1,368 males and 240 of 607 females, aged 10–90[+], had asbestos-like bodies in their lungs. Such bodies may or may not be carcinogenic.

If they are carcinogenic, are the dosage levels high enough to cause lung cancer or other cancers? The chance of developing a malignancy depends upon lifetime exposure, the amount of the asbestos that is swallowed, the length of the fibers, and the type of asbestos and seems to be affected by other activities, such as cigarette smoking.

Much of the research on asbestos in the environment has focused on obvious sources of asbestos, such as automobile repair shops, construction sites, and toll booths. For example, air concentrations 100 times higher than background levels are occasionally found near construction sites (Nicholson and Pundsack, 1973). At toll booths in New York City, Bruckman and Rubino (1975) found concentrations three to five times higher than normal background rates.

During the late 1970's, Rohl, Langer, and Selikoff (1977) and Carter (1977) discovered asbestos in the crushed serpentine rock used to pave roads and other surfaces in Montgomery County, Maryland. Initial air samples taken in the vicinity of roads paved with this material showed the concentration of the carcinogen chrysotile asbestos to be 1,000 times greater than that typical for urban ambient air in the United States. Water supplies and soil would also be contaminated by runoff. High levels of asbestos in the ambient environment were also discovered in a recreational area in San Benito County, California (Cooper et al., 1979). These findings have led scientists to begin a systematic check of quarries around the United States for other asbestos veins.

Another concern is that an air pollutant, studied in isolation in the laboratory, may not itself produce cancer, but, combined with some other constituent of the air, it may be activated as a carcinogen or it may act synergistically to potentiate the effect of another constitutent that is or could be carcinogenic. With respect to the two classes of substances that have just been discussed, Selikoff (1968, 1977) and Miller et al. (1965) have reported that such hydrocarbons as benzo(a)pyrene adhere to asbestos fibers, thereby increasing the carcinogenic potential of both. Such synergism would be particularly threatening in urban areas because of the great variety of substances present.

Although organic chemicals and asbestos have attracted most of the attention, other pollutants are being studied. Scientists estimate that 6 percent of the vinyl chloride, which causes liver cancer, used in the production of plastics, is lost to the environment, with 4.5 percent being vented into the air. Furthermore, vinyl chloride was formerly used as a propellant in pesticide sprays, and the U.S. Environmental Protection Agency (EPA) has reported railroad and truck accidents involving tanks of vinyl chloride (Office of Toxic Substances, U.S. EPA, 1974). Other substances may be equally dangerous. A preliminary study indicated that the lung cancer mortality rate for census tracts adjacent to an arsenical pesticide plant in Baltimore was three to four times higher than in control tracts, the rates decreasing with increasing distance from the plant (American Public Health Association, 1976). Sawicki's careful review of the literature (1977) for the International Agency for Research on Cancer demonstrates that many other carcinogens and cocarcinogens can be found in the urban atmosphere, some in high concentrations.

The newest issue in the controversy over air pollution and lung cancer is the possible role of indoor air pollution. Some scientists have argued that this recently discovered component of air pollution is far more significant than outdoor air pollution because people spend about 70 to 80 percent of their time indoors. Initial sampling of indoor air has disclosed high concentrations of chemicals derived from cigarette smoke, gas stoves, insulation, and radioactive substances found in the surrounding bedrock (see, e.g., Repace and Lowrey, 1980).

The main arguments for a relationship between urban air pollution and respiratory cancer thus are that urban air contains substances known to cause cancer (cigarette smoke, industrial pollutants, motor vehicle exhaust, construction materials) and that an urban excess of lung cancer cannot always be attributed solely to cigarette smoking and occupational exposure. The arguments against an urban air pollution factor are that the cities with the worst quality air do not necessarily have the highest lung cancer rates, that the much higher lung cancer rates in men than in women seem to be associated with their smoking habits and occupations, and that lifetime residents of urban areas do not have much higher lung cancer rates than have migrants. Although some air pollutants are considered carcinogenic, particularly in urban/industrial areas, pollutants from cigarette smoke and industrial processes are considered much more important in the etiology of respiratory cancer.

DIGESTIVE SYSTEM

In contrast to respiratory system cancers, relatively few studies have shown an excess of digestive system cancers in urban areas. The reported excess of large bowel cancer in urban areas has almost always been attributed to one factor: diet. Calling cancer of the large bowel "an epidemiologic jigsaw puzzle," Burkitt (1975) noted the close association of the disease with economic development and urbanization. He theorized that the absence of wholemeal bread and the prevalence of fats in the Western diet are key factors. Burkitt observed that most of the cancers occur in the part of the bowel where feces tend to accumulate, indicating an important association between prolonged residence of feces and the activity of bowel bacteria. Other investigators have also emphasized the role of diet (Burdette, 1975; Walker and Burkitt, 1976; Newberne and Rogers, 1976; Graham, 1979). Norden (1979), in experiments with mice, found that chronic energy deficiency due to dietary restrictions inhibits the formation of many tumors; he also observed that epidemiological studies indicate that obesity can increase the risk of developing tumors, particularly in the intestinal tract, liver, gallbladder, breast, uterus, and genitourinary tract. The National Academy of Sciences (1982) has made dietary suggestions that include reducing the intake of fats, of salt-cured, salt-pickled, and smoked foods, and alcohol, and increasing the intake of whole-grain cereals, fruits, and vegetables. These should have an effect on digestive organ cancers, especially the intestines, stomach, and esophagus.

Kagawa (1978) has provided an interesting historical perspective on the relationship between Western dietary practices and high colon cancer rates. Although colon cancer rates in Japan were once among the lowest in the world, the incidence of colon and breast cancer increased two- to threefold between 1950 and 1975, and during this same period the Japanese diet appears to have changed dramatically. The intake of milk increased 15-fold; of meat, poultry, and eggs seven and one-half fold; and of fat sixfold. During the same period, rice consumption decreased 70 percent, potato consumption 50 percent, and barley consumption about 3 percent. These dietary changes and the increase in colon cancer are reported to be most characteristic of nonfarmers, city dwellers, and the wealthy. Although Kagawa attributes the urban excess of colon cancer in Japan to dietary changes, an epidemiological study in Japan of 588 patients with colorectal cancer and 1,176 controls found only weak (statistically insignificant) positive associations with social class and urbanization (Haenszel et al., 1980).

In order to investigate the dietary hypotheses and obtain further leads, the National Cancer Institute (Blot et al., 1976, 1977a) analyzed the 1950–69 U.S. cancer mortality data on cancer of the large intestine. White male and white female rates were highest in the Northeast and lowest in the South. The highest rates were correlated with urbanization, high socioeconomic status, and ethnicity (Irish, German, Czechoslovakian). Rectal cancer had similar correlates and a similar geographical distribution. However, findings from case control studies in the United States have not been consistent in showing a link between colorectal cancers and diet (Jain, Miller, and Howe, 1979; Sorenson and Lyon, 1979; Dales et al., 1979).

Less attention has been paid to the relationship of urbanization to other digestive system cancers. Pointing to higher pancreatic cancer death rates in urban areas of the United States and in other Western nations, Fraumeni (1975) suggested cigarette smoking, diabetes, alcohol consumption, chemical agents in the workplace, and dietary fat as major factors. Krain (1970) and Blot et al. (1978) stressed the contribution of cigarette smoking to the increasing rate of this type of cancer, while minimizing genetic factors and air pollution. Peacock (1976) suggested a role for psychological stress factors in pancreatic cancer, a possibility supported by research demonstrating that stress influences tumor growth in animals (Sklar and Anisman, 1979).

Cancers of the stomach and esophagus have also been linked to diet and urbanization. Adelstein (1972) reported higher rates of stomach cancer in urban British males. Wynder et al. (1963a) observed that stomach cancer patients have a low intake of fresh vegetables and fruits and a high intake of starchy foods, as well as a higher consumption of heavily salted, smoked, and charcoal-broiled food. Low socioeconomic status is a common thread running through the literature on stomach cancer.

Newberne and Rogers (1976) and Mettlin et al. (1980) have proposed the following dietary factors to be of etiological importance in esophageal cancer: high alcohol consumption and decreased milk, egg, and green leafy vegetable consumption. Heavy smoking has also been mentioned. In a nationwide study of U.S. counties with shipyard industries, Blot et al. (1979) found high rates of esophageal cancer in the urban Northeast.

Overall, since nutrition appears to be the strongest factor in digestive system cancer and since urban cancer mortality rates of intestinal, esophageal, and pancreatic cancer are higher than rural rates, it follows that there may be urban and rural dietary differences that, when combined with other risks more prevalent in the urban than the rural environment (e.g., smoking, alcohol consumption), increase the risk of the city dweller contracting cancer of the digestive tract.

FEMALE BREAST CANCER

The large epidemiological literature on female breast cancer deals mainly with genetic, hormonal, and dietary factors (Hems, 1970; Seidman, 1971; Cole, 1974; King et al., 1979; Bain et al., 1980; Ottman and King, 1980). A genetic component is suggested by the consistent familial aggregation of female breast cancer.

Two forms of female breast cancer have been hypothesized: early and late. The first tends to be associated with early age at menopause and excess ovarian estrogen (Valaoras et al., 1969; Yuasa and MacMahon, 1970). The second occurs among more elderly women and is associated with obesity, hypertension, and diabetes, which are related to hormonal disturbances. Both types are more frequently found in relatively high socioeconomic status populations (Carrol et al., 1968; Wynder, 1969; Fraumeni et al., 1969; Hems, 1970).

Female breast cancer rates have been reported to be higher in urban areas than in rural areas (Vakil and Morgan, 1973; Ericson et al., 1976). In the most comprehensive geographical study of female breast cancer, Blot, Fraumeni, and Stone (1977b) analyzed the distribution of the disease across the United States at the county scale. For women aged 20–44, the strongest correlates in order of importance were urbanization, low birth rates, and high ovarian cancer mortality rates. For women over 55 years old, urbanization was the fifth highest correlate. Exceeding it in importance were regional location (the Northeast had the highest rates), high socioeconomic status, ethnicity (German heritage), and high colon cancer mortality rates. Other significant correlates were high ovarian cancer mortality rates, low birth rates, and Scandinavian heritage. These statistical results were consistent with what the authors had expected on the basis of previous studies, but the findings did not fully explain the observed geographical variations of female breast cancer. Even more than the urban diet/urban digestive cancer link, the observed urban excess in female breast cancer remains unexplained.

URINARY SYSTEM

Until the mid 1970's, urinary system cancer had been attributed to industrial exposures, smoking, and diet (e.g., Wynder, Onderdonk, and Mantel, 1963b). At least industrial exposure and smoking should lead to an urban excess of these urinary system cancers. More recently, studies have suggested that water supplies in some urban areas may play a role.

Industrial exposures have shown the most consistent relation to male urinary system cancer in studies of aniline dye workers, shoe repairers, leather workers, painters, hairdressers, some textile workers, coal miners, and plumbers. Blot et al. (1977a) reported high rates of male bladder cancer mortality in industrial counties of the United States, but other studies did not show similar relationships. Ohno and Aoki (1977), for example, found no association between Japanese bladder cancer mortality rates and industrial activity or urbanization for the period 1947–74.

Hypotheses concerning diet have focused on the possible causative roles of saccharin, sodium cyclamate, and coffee and the beneficial roles of vitamins A, C, and B_6 (Mettlin et al., 1979; Mettlin and Graham, 1979; Friedell, Greenfield, and Cohen, 1979). Simon, Yen, and Cole (1975) reported a significant difference in urinary system cancer rates between coffee drinkers and non-coffee–drinkers, but since no dose-response relationship was observed, they concluded that there was no causal relationship. Kantor (1977) found consistently higher renal cancer mortality rates in urban areas of the United States and, while noting that the disease is relatively rare, he attributed the urban excess to radiation, chemical agents, and hormonal factors. It is hard to explain why hormonal factors in cities should be different from those in the country.

To the extent that cigarette smoking, certain industrial processes, and some dietary habits characterize urban areas, one would expect higher urban rates of bladder and perhaps other urinary system cancer based on studies done before the mid 1970's. Until then, little attention had been given to the possibility of an association between the water supplies of urban areas and cancer of the urinary and digestive systems. In 1974, Robert Harris announced that white male cancer mortality rates for all sites and for the urinary and gastrointestinal systems were very high in those parishes (counties) of Louisiana, especially New Orleans, where people drink Mississippi River water. These findings led to the passage of the Safe Drinking Water Act (Rogers, 1978) and stimulated much research.

The role of water supplies in explaining higher urban cancer mortality rates is based on the assumption that the surface water supplies of some cities are contaminated by industrial waste, effluent from public sewerage treatment plants, and runoff, especially from landfills. Trihalomethanes (chloroform, dichlorobromomethane, dibromochloromethane, and bromform light chlorinated hydrocarbons) have also received wide attention because of their carcinogenic potential and their presence in drinking water, presumably due to its chlorination (Hileman, 1982).

Since Harris' findings, more than 20 studies have explored the relationship between cancer rates and water as a risk factor (Wilkins et al., 1979; Schneiderman, 1978; U.S. National Research Council, 1978). Most have correlated age-, race-, and sex-standardized cancer mortality rates for the white population with water source or water quality data, while controlling for such confounding factors as industrialization, socioeconomic status, and urbanization. The results are inconclusive, although there seems to be an otherwise unexplained relationship between gastrointestinal and urinary system cancer mortality rates and surface water supply sources and/or trihalomethanes in water. Recently, a few case control studies have sought to isolate the impact of water supply by controlling for all major factors except water source and treatment.

In the Louisiana studies, Harris (1974) correlated white male total, urinary, and gastrointestinal cancer mortality rates by parish (county) in

Louisiana with percentage of water derived from the Mississippi River. His control variables were urbanization, occupation, population density, and median family income. From the analysis, he estimated that there were over 50 premature white male deaths annually that could be attributed to the water supply.

Subsequently, Page, Harris, and Epstein (1976) expanded the population to include white female and nonwhite males and females as well as white males, and a larger number of body sites, and they addressed the issues of population migration and missing data (e.g., information on diet). Their results were essentially the same as those in the original study.

Despite the influence of Harris' studies, they have been criticized by other investigators. Tarone and Gart (1975) re-examined the Louisiana data for white females, and nonwhite males and females; they added elevation above sea level as a factor and employed weighted regression as an analytic method. Elevation was found to be more important than the water supply. Perhaps elevation was related to urinary cancer because it acts as a surrogate for such factors as geological structure and land use. Or perhaps the elevation-cancer relationship was spurious. Whichever, the elevation-cancer link implied that the water supply/cancer link was, at best, a weaker factor than Harris had concluded.

DeRouen and Diem (1975, 1977) criticized the Harris studies indicating that there was "considerable inconsistency in the evidence supporting the hypothesis of the water as a causative agent, [that there were] contradictions in the theory that were overlooked," and that it was "possible to use the same data to support a different hypothesis." For example, nine of the 45 counties in the United States with the highest total white male cancer rates during 1950–69 are found in Louisiana. Four of the nine, however, are rural counties not using the Mississippi River as a source of their public potable water supply. Despite their criticisms of Harris' work, DeRouen and Diem found positive associations between urinary and gastrointestinal cancer and water supply. Rather than taking the large leap from ecological clues to excess mortality estimates, however, they argued that the results were not consistent enough to consider drinking water an important factor.

The controversial Louisiana studies are ongoing. In the initial results of a case control study of selected parishes, 1965–75, Gottlieb et al. (1979) found an increased risk (over 1.6) of kidney, rectal, and to a lesser extent, liver cancer for those consuming surface and chlorinated water and for persons who had lived more than ten years in the parish. Subsequent results point to an excess of rectal cancer (Gottlieb et al., 1980).

Other studies have been conducted in the Ohio River Basin (Buncher, 1975; Kuzma, Kuzma, and Buncher, 1977; Harris et al., 1977; Salg, 1977), Los Angeles County (Mah, Spivey, and Sloss, 1977), upper New York State (Alavanja, Goldstein, and Susser, 1976), Iowa cities with a population of at least 25,000 (Bean, Isacson, and Hausler, 1979), Pittsburgh (Carlson and Andelman, 1977), Hagerstown in Washington County, Maryland (Kruse, 1977), and North Carolina (Struba and Shy, 1979). Most, though not all, of these investigations have reported some statistically significant correlations between water supply and urinary and digestive system cancer rates.

Another group of studies focused on the organic chemical/cancer link in selected counties throughout the United States rather than within a specific region of the country. Cantor et al. (1978) correlated trihalomethane concentrations with age-standardized white male and female cancer mortality rates for 1968–71 in 923 counties, all of which were at least 50 percent urban in 1970. Their analysis was controlled for median school years completed, ethnicity, population change, percentage urban, presence of industry, and region of the United States. They found statistically significant correlations between trihalomethanes and bladder cancer. Hogan et al. (1979) performed similar analyses and found significant associations between chloroform and bladder and rectal/intestine cancer mortality rates, but reiterated many of the five methodological criticisms of correlation studies from county scale data presented earlier in the chapter (e.g., ecological fallacy, confounding variables, long latency period).

In addition to organic contaminants, other substances have been investigated. Because of the concern over asbestos discharges into Lake Superior from taconite ore waste since 1955, water quality and cancer rates have been studied in Duluth, Minnesota. The water analyses revealed high concentrations of asbestos-like fibers in Duluth's tapwater (Cook et al., 1974), but preliminary epidemiological studies comparing Duluth with Minneapolis and St. Paul did not disclose any clear differences in cancer rates (Levy et al., 1976). On the other hand, Kanarek et al. (1980) found positive associations between asbestos in drinking water and cancer in the San Francisco Bay Area. There is chrysotile asbestos in Bay Area serpentine bedrock. Age-adjusted and sex- and race-specific cancer incidence ratios for 1969–71 were computed for 722 census tracts in the region. These were compared with average numbers of chrysotile asbestos fibers in 353 samples taken during 1977–78. After controlling for socio-economic status, occupational exposure, ethnicity, mobility, and marital status, the authors found a statistically significant association between asbestos concentration in drinking water and white male and female peritoneal and stomach cancer and white male lung and white female gallbladder, pancreas, esophagus, pleura, and kidney cancer.

Asbestos is not the only mineral of concern. Berg and Burbank (1972) correlated data on 34 types of cancer gathered over 18 years in 15 river basins with data on eight inorganic metals: arsenic, aluminum, nickel, iron, beryllium, cadmium, cobalt, and lead. The results are to be viewed with caution because their correlations do not always fit the expected pattern. For example, as expected from animal studies, beryllium was associated with bone cancer and lead with leukemia and bowel and kidney cancer. Chromium, a known lung and nasal cavity carcinogen, however, did not correlate with cancer of any body site.

To establish credibility for a connection between urban water supplies and cancer, carcinogens must, of course, be found in urban water supplies. Since the late 1960's, they have been detected, though usually in raw untreated water (see Wilkins et al., 1979 for a brief review). Although these studies were informative to specialists, many scientists, as well as the public, were alarmed when the U.S. Environmental Protection Agency sampling programs required by the Safe Drinking Water Act demonstrated the presence of carcinogenic substances in city water supplies throughout the United States. In November, 1975, the Office of Toxic Substances of the EPA reported small amounts of 253 different organic chemicals in urban drinking water supplies. The majority of the chemicals had not been tested for carcinogenicity, but some were classified as carcinogens or as suspected carcinogens by animal testing. Moreover, the 253 chemicals represented only a small fraction of the total organic content of the samples, and therefore other substances of equal or greater danger may have been present in the drinking water samples. The highest concentration was 366 μg/liter for the carcinogen chloroform; the lowest was 0.001 μg/liter for the pesticide dieldrin. Such inorganic chemicals as arsenic, beryllium, cadmium, chromium, nickel, and selenium, and nitrogenous compounds, were also found by these surveys. Asbestos appeared in less than one-half of the city drinking supplies. Traces of radionuclides were found in most.

Perhaps the most alarming finding of these initial surveys was the fact that data gathered from 50 cities showed a significant correlation between concentrations of chloroform in water samples, presumably due to chlorination of drinking water, and total cancer rates for the period 1969–71. A parallel study of 43 cities, however, showed no relationship between chloroform concentrations and cancer mortality and when the shortcomings of data and methods mentioned throughout this chapter were also considered, the initial positive results were severely weakened.

As a followup, the EPA made a thorough literature review for evidence of the presence of organic chemicals in water and found 5,700 literature entries mentioning 1,259 different compounds (Shackelford and Keith, 1977). Mass spectrometic and gas chromatographic methods can now detect concentrations of these compounds in parts per trillion, which make possible efforts to measure such compounds. The State of New Jersey is pioneering many of these efforts through its association with the National Cancer Institute's cancer/water quality studies (Burke and Tucker, 1978; Greenberg and Page, 1981; Page, 1981; Greenberg et al., 1982). Therefore, better evidence of the extent of chemical contamination of drinking water supplies should become available within the next five years.

To conclude this section, there are two main arguments for an urban water pollution factor: (1) city water supplies contain known and suspected carcinogens (2) and there is an urban excess of urinary and some digestive system cancers that has been statistically associated with water quality and source of drinking water, even after controlling for some confounding factors. The argument against an urban water pollution factor is that too few case-control studies and too few thorough surveys of carcinogens in drinking water have been done to warrant a conclusion that the investigations that do show an association between cancer and water supply offer anything more than a strong warning that the chemical content of drinking water should be continuously monitored.

SUMMARY

There is an urban excess of respiratory and some digestive and urinary system and female breast cancer, which poses the question of how to explain this excess. The only "urban factor" consistently mentioned in the literature is air pollution, but it is not too great a leap to include water pollution. Overall, when limited to pollution, the contribution of the urban factor to the urban excess of cancer is of limited importance.

It would be useful, however, to include in the urban factor the personal, occupational, and environmental risk factors that have come together in urban areas and that are thought to contribute to high cancer mortality rates. This would place urban cancer excess in a better historical context. First, large cities developed, with populations who exhibited the following characteristics: high levels of cigarette smoking and drinking; a poor diet from the perspective of cancer prevention; cultural and genetic traits of European origin that predisposed to higher risks for certain types of cancer; factory work that exposed individuals to dangerous substances; and residence in areas where the quality of the air and water left much to be desired. Then, the tide of urbanization and industrialization spread from the central cities to the suburbs and from the Northeast and the Great Lakes Region to other regions, but at the same time the European influence on American culture gradually declined. With these developments one would expect a shift from higher to lower differences in cancer rates between the cities, suburbs and old industrial belt, on the one hand, and the rest of the United States, on the other.

REFERENCES

Adelstein, A. 1972. "Occupational mortality: Cancer." *Ann. Occup. Hyg.* 15: 53–57.

Alavanja, M., Goldstein, I., and Susser, M. 1976. Report of case control study of cancer deaths in four selected New York counties in relation to drinking water chlorination, submitted to U.S.E.P.A. Cincinnati, Oh.: Health Effects Research Laboratory.

American Public Health Association. 1976. For Office of Toxic Substances, USEPA, Epidemiology Studies, Task 1, Phase 1: Pilot Study of Cancer Mortality Near an Arsenical Pesticide Plant in Baltimore, Springfield, Virginia: National Technical Information Service (NTIS).

Asahima, S., Andrea, J., Carmel, A., Arnold, E., Bishop, Y., Joshi, S., Coffin, D., and Epstein, S.S. 1972. "Carcinogenicity of organic fractions of particulate pollutants collected in New York City and administered subcutaneously to infant mice." *Cancer Res.* 32 (10): 2263–2268.

Bain, C., Speizer, F., Rosner, B., Belanger, C., and Henneken, C. 1980. "Family history of breast cancer as a risk indicator for the disease." *Amer. J. Epidemiol.* 111: 301–308.

Bean, J., Isacson, P., and Hausler, W. Jr. 1979. Drinking water characteristics and stomach cancer risk. Paper presented at Society for Epidemiologic Research, New Haven, Connecticut, June 13, 1979.

Berg, J., and Burbank, F. 1972. "Correlations between carcinogenic trace metals in water supplies and cancer mortality." *Ann. N. Y. Acad. Sci.* 199: 249–264.

Blot, W., Fraumeni, J., Stone, B.J., and McKay, F. 1976. "Geographic patterns of large bowel cancer in the United States." *J. Natl. Can. Inst.* 57: 1225–1231.

Blot, W.J., Mason, T.J., Hoover, R., and Fraumeni, J.F. 1977a. "Cancer by county: Etiologic implications." In H. Hiatt, J.D. Watson, and J.A. Winsten, eds. *Proceedings of the Cold Spring Harbor Conferences on Cell Proliferation, Origins of Human Cancer,* Cold Spring Harbor, New York, 1977, pp. 21–32.

Blot, W.J., Fraumeni, J.F., and Stone, B.J. 1977b. "Geographic patterns of breast cancer in the United States." *J. Natl. Can. Inst.* 59 (5): 1407–1411.

Blot, W., Fraumeni, J., and Stone, B. 1978. "Geographic correlates of pancreas cancer in the United States." *Cancer* 42: 373–380.

Blot, W.J., Stone, B.J., Fraumeni, J.F., and Morris, L.E. 1979. "Cancer mortality in U.S. counties with shipyard industries during World War II." *Environ. Res.* 18 (2): 281–290.

Bruckman, L., and Rubino, R. 1975. "Asbestos: Rational behind a proposed air quality standard." *J. Air Pollut. Contr. Assoc.* 25: 1207–1215.

Buell, P., and Dunn, J. 1967. "Relative impact of smoking and air pollution on lung cancer." *Arch. Environ. Health* 15: 291–297.

Buell, P., Dunn, J., and Breslow, I. 1967. "Cancer of the lung and Los Angeles type air pollution: Prospective study." *Cancer* 20: 2139–2147.

Buncher, C. 1975. "Cincinnati drinking water: An epidemiological study of cancer rates." Division of Epidemiology and Biostatistics, University of Cincinnati, University of Cincinnati Medical Center.

Burdette, W. 1975. "Geoprevalence and etiology of cancer." *Bull. Soc. Int. Chir.* 34: 345–354.

Burke, T., and Tucker, R. 1978. "A preliminary report on the findings of the state ground water monitoring project," Trenton, New Jersey: Program on environmental cancer and toxic substances. State of New Jersey, Department of Environmental Protection.

Burkitt, D. 1975. "Large-bowel cancer: An epidemiological jigsaw puzzle." *J. Natl. Can. Inst.* 54: 3–6.

Cantor, K., Hoover, R., Mason, T., and McCabe, L. 1978. "Association of cancer mortality with halomethanes in drinking water." *J. Natl. Can. Inst.* 61: 979–985.

Carlson, W., and Andelman, J. 1977. "Environmental influences on Cancer Morbidity in the Pittsburgh Region." Report to U.S.E.P.A., Cincinnati, Ohio, 1977.

Carrol, K., Gammal, E., and Plunkett, E. 1968. "Dietary fat and mammary cancer." *Canad. Med. Assoc. J.* 98: 590.

Carter, L. 1977. "Asbestos: Trouble in the air from Maryland rock quarry." *Science* 196:237–240.

Cole, P. 1974. "Epidemiology of human breast cancer." *J. Invest. Dermatol.* 63 (1): 133–137.

Cook, P., Glass, G., and Tucker, J. 1974. "Asbestiform amphibole materials: Detection and measurement of high concentrations in municipal water supplies." *Science* 185: 853–855.

Cooper, W. 1967. "Asbestos as a hazard to health." *Arch. Environ. Health* 15: 285–290.

Cooper, W., Murchio, J., Popendorf, W., and Wenk, H. 1979. "Chrysotile asbestos in a California recreational area." *Science* 26: 685–688.

Council on Environmental Quality. 1976. *The Sixth Annual Report on Environmental Quality*, Washington, D.C.: U.S. Government Printing Office.

Dales, L., Friedman, G., Ury, H., Grossman, S., and Williams, S. 1979. "A case-control study of the relationship of diet and other traits to colorectal cancer in American blacks." *Amer. J. Epidemiol.* 109:132–144.

Dean, G. 1961. "Lung cancer among White South Africans." *Brit. Med. J.* 2: 852, 1599.

DeRouen, T., and Diem, J. 1975. "The New Orleans drinking water controversy. A statistical perspective." *Amer. J. Public Health* 65: 1060–1062.

DeRouen, T., and Diem, J. 1977. "Relationship between cancer mortality in Louisiana drinking water sources and other possible causative agents." In H. Hiatt, J. Watson, and J. Winsten, eds. See Blot et al. (1977a) for full reference.

DeWaard, R. 1969. "The epidemiology of breast cancer; Review and prospects." *Internat. J. Can.* 4: 577.

Doll, R. 1963. "Investigation into cigarette smoking and atmospheric pollution in the aetiology of lung Cancer" *Methods Inform. Med.* 2: 13–19.

Doll, R., and Hill, A.B. 1964. "Mortality in relation to smoking: Ten years' observation of British doctors." *Brit. Med. J.* 1: 1460–1467.

Doll, R. 1970. "Practical steps toward the prevention of bronchial carcinoma." *Scot. Med. J.* 15: 433–447.

Doll, R., and Peto, R. 1981. *The Causes of Cancer,* New York: Oxford University Press.

Eastcott, D. 1961. In S. Farber and R. Wilson, eds. *The Air We Breathe.* Springfield, Ill. Thomas.

Epstein, S., Andrea, S., Mantel, N., Sawicki, E., Stanley, T., and Tabor, E. 1966. "Carcinogenicity of organic particulate pollutants in urban air after administration of trace quantities to neonatal mice." *Nature* 212: 1305–1307.

Ericson, J. L., Karnstrom, L., Mattson, B., and Willgran, J. 1976. "The incidence of breast and cervix cancer in the Swedish population." In H. Bostrom and T. Larsson, eds. *Health Control in the Detection of Cancer,* Stockholm: Almquvist & Wiksell International.

Falk, J., Kotin, P., and Mehler, A. 1964. "Polycyclic hydrocarbons as carcinogens for man." *Arch. Environ. Health.* 8: 721–730.

Fox, A.J., Lindars, D.C., and Owen, R. 1974. "A survey of occupational cancer in the rubber and cablemaking industries: Results of five-year analysis, 1967–1971." *Brit. J. Ind. Med.* 21 (2): 140–151.

Fraumeni, J. Jr., Lloyd, J., Smith, E., and Wagoner, J. 1969. "Cancer mortality among nuns: Role of marital status in etiology of neoplastic disease in women." *J. Natl. Can. Inst.* 42: 455.

Fraumeni, J.F. 1975. "Cancers of the pancreas and biliary tract: Epidemiological considerations." *Can. Res.* 35 (11:2): 3437–3446.

Friedell, G., Greenfield, R., and Cohen, S. 1979. "Nutritional factors that may be involved in cancer of the bladder." *Nutr. Can.* 1: 82–88.

Gerrard, M., from Kireeva, I., and Yanysheva, N. 1970. *Loading of Atmospheric Air with Carcinogenic Aromatic Hydrocarbons from Petroleum Processing Installations,* Springfield, Virginia: NTIS.

Gordon, R.J., Bryan, R.J., Rhim, J.S., Demoise, C., Wolford, R.G., Freeman, A.E., and Huebner, R.J. 1973. "Transformation of rat and mouse embryo cells by a new class of carcinogenic compounds isolated from particles in city air." *Intl. J. Can.* 12 (1): 223–232.

Gottlieb, M., Shear, C., Seale, D., and Stedman, R. 1979. "Cancer mortality in Louisiana parishes and potable water sources." Paper presented at Society for Epidemiologic Research, New Haven, Connecticut, June 14, 1979.

Gottlieb, M., Carr, J., and Morris, D. 1980. "The epidemiology of colo-rectal cancer associated with drinking water source." Paper presented at Thirteenth Annual Meeting of the Society for Epidemiologic Research, Minneapolis: June 18, 1980.

Graham, S. 1979. "Diet and colon cancer." *Amer. J. Epidemiol.* 109: 1–20.

Greenberg, M., and Page, G.W. 1981. "Planning with great uncertainty: A review and case study of the safe drinking water controversy." *Socio-Econ. Plan. Sci.* 15: 65–74.

Greenberg, M., Anderson, R., Keene, J., Kennedy, A., Page, G.W., and Schowgurow, S. 1982. "Empirical test of the association between gross contamination of wells with toxic substances and surrounding land use." *Environ. Sci. Technol.* 16: 14–19.

Greenburg, L., Field, F., Reed, J., and Glasser, M. 1967. "Air pollution and cancer mortality." *Arch. Environ. Health* 15: 356–361.

Grover, P. 1973. "How polycyclic hydrocarbons cause cancer." *New Sci.* 58 (850): 685–687.

Haenszel, W., Loveland, D., and Sirken, M. 1962. "Lung cancer mortality as related to residence and smoking histories. I. White males." *J. Natl. Can. Inst.* 28: 947–1001.

Haenszel, W., and Taeuber, K. 1964. "Lung cancer mortality as related to residence and smoking histories. II. White females." *J. Natl. Can. Inst.* 32: 803–838.

Haenszel, W., Locke, F., and Segi, M. 1980. "A case control study of large bowel cancer in Japan." *J. Natl. Can. Inst.* 64: 17–22.

Hagerstrand, I., and Seifert, B. 1973. "Asbestos bodies and pleural plaques in human lungs at necropsy." *Acta Pathol. Microbiol. Scand.* 81 (4): 457–460.

Hagstrom, R., Sprague, H., and Landau, E. 1967. "The Nashville air pollution study: VII. Mortality from cancer in relation to air pollution." *Arch. Environ. Health.* 15: 237–248.

Hallenbeck, W., Hesse, C., Chen, E., Pantel-Mandlik, K., and Wolff, A. 1977. *Asbestos in Potable Water,* Springfield, Va: NTIS.

Hammond, E., and Horn, D. 1958. "Smoking and death rates." *J. Amer. Med. Assoc.* 166: 1294–1308.

Harris, R. 1974. "Implications of cancer-causing substances in Mississippi River water." Washington, D.C.: Environmental Defense Fund.

Harris, R., Page, T., and Reiches, N. 1977. "Carcinogenic hazards of organic chemicals in drinking water." In H. Hiatt, J. Watson, and J. Winsten, eds. For full reference see Blot (1977a).

Hems, G. 1970. "Epidemiological characteristics of breast cancer in middle and late age." *Brit. J. Can.* 24: 226.

Higgins, I.T. 1976. "Smoking and cancer." *Amer. J. Public Health* 66 (2): 159–161.

Hileman, B. 1982. "The chlorination question." *Environm. Sci. Technol.* 16: 15A–18A.

Hogan, M., Chi, P., and Hoel, D. 1979. "Association between chloroform levels in finished drinking water supplies and various site-specific cancer mortality rates." *J. Environ. Pathol. Toxicol.* 2: 873–887.

Hueper, W. 1966. *Occupational and Environmental Cancers of the Respiratory System.* New York: Springer-Verlag.

Jain, M., Miller, A., and Howe, G. 1979. "Diet and bowel cancer." Paper presented at twelfth annual meeting, Society for Epidemiologic Research, New Haven, Conn., June 13, 1979.

Kagawa, Y. 1978. "Impact of Westernization on the nutrition of Japanese: Changes in physique, cancer longevity and centenarians." *Prev. Med.* 7: 205–217.

Kanarek, M., Conforti, P., Jackson, L., Cooper, R., and Murchio, J. 1980. "Asbestos in drinking water and cancer incidence in the San Francisco Bay Area." *Amer. J. Epidemiol.* 112: 54–72.

Kantor, A. 1977. "Current concepts in the epidemiology and etiology of primary renal cell carcinoma." *J. Urol.* 117 (4): 415–417.

King, M.C., Elston, R., Go, R., and Lynch, H. 1979. "Genetic analysis of breast cancer in families." Paper presented at twelfth annual meeting, Society for Epidemiologic Research, New Haven, Conn., June 13, 1979.

Koshal, R., and Koshal, M. 1973. "Environments and Urban Mortality: An Econometric Approach." *Environ. Pollut.* 4: 247–259.

Krain, L.S. 1970. "The rising incidence of carcinoma of the pancreas: An epidemiologic appraisal." *Amer. J. Gastroenterol.* 54 (5): 500–507.

Kruse, C. 1977. "Chlorination of public water supplies and cancer, Preliminary Report for Washington County, Maryland." Report to U.S. EPA., Cincinnati, Ohio.

Kuzma, R., Kuzma, C., and Buncher, C. 1977. "Ohio drinking water source and cancer rates." *Amer. J. Public Health* 67: 725–729.

Lave, L., and Seskin, E. 1970. "Air pollution and human health." *Science* 169: 723–733.

Lave, L., and Seskin, E. 1971. "Health and air pollution: The effect of occupation mix." *Swed. J. Econ.* 73: 76–95.

Lave, L., and Seskin, E. 1972. "Air pollution, climate, and home heating: Their effects on U.S. mortality rates." *Amer. J. Public Health* 62: 909–916.

Lawther, P. 1965. "Air pollution," *N.Y. Acad. Med.* 41: 214–217.

Levin, D. et al., 1974. *Cancer Rates and Risks,* Washington, D.C.: DHEW Pub. No. NIH 75-691.

Levin, M. 1960. "Cancer incidence in urban and rural areas of New York State." *J. Natl. Can. Inst.* 24: 1243–1257.

Levy, B., Sigurdson, E., Mandel, J., Laudon, E., and Pearson, J. 1976. "Investigating possible effects of asbestos in city water: Surveillance of gastrointestinal cancer incidence in Duluth, Minnesota." *Amer. J. Epidemiol.* 103: 362–368.

Mabuchi, K., Lilienfeld, A.M., and Snell, L.M. 1979. "Lung cancer among pesticide workers exposed to inorganic arsenicals." *Arch. Environ. Health* (5): 312–320.

MacDonald, G. and Schwing, R. 1973. "Instabilities of regression estimates relating air pollution to mortality." *Technometrics* 15: 463–481.

Mah, R., Spivey, G., and Sloss, E. 1977. Cancer and chlorinated drinking water. Report to U.S. EPA, Cincinnati, OH.

Manos, N. and Fisher, G. 1959. "An index of air pollution and its relation to health." *J. Air Pollut. Contr. Assoc.* 9: 5–11.

Marmor, M. 1978. "Air pollution and cancer in Houston, Texas: A causal relationship." *J. Amer. Med. Wom. Assoc.* 33 (6): 275–278.

Menck, H., Casagrande, J., and Henderson, B. 1974. "Industrial air pollution: Possible effect on lung cancer." *Science* 183: 210–212.

Mettlin, C., Graham, S., and Rzepka, T. 1979. "Vitamin A in human bladder cancer." Paper presented at twelfth annual meeting, Society for Epidemiologic Research, New Haven, Conn., June 13, 1979.

Mettlin, C. and Graham, S. 1979. "Dietary factors in human bladder cancer." *Amer. J. Epidemiol.* 110: 255–263.

Mettlin, C., Graham, S., Priore, R., and Swanson, M. 1980. "Diet and cancer of the esophagus." Paper presented at Thirteenth Annual Meeting of the Society for Epidemiologic Research, Minneapolis, June 18, 1980.

Miller, L., Smith, W., and Berlinger, S. 1965. "Tests for effect of asbestos on benzo(a)pyrene carcinogenesis in the respiratory tract." *Ann. N.Y. Acad. Sci.* 132: 489–500.

National Academy of Sciences. 1971. *Asbestos: The Need For and Feasibility of Air Pollution Controls,* Springfield, Va: NTIS.

National Academy of Sciences. 1972. *Particulate Polycyclic Organic Matter.* Springfield, Va: NTIS.

National Academy of Sciences, Conference on Health Effects of Air Pollution, October 3–5, 1973. Summary of Proceedings. Prepared for the Committee on Public Works of the U.S. Senate, 1973.

National Academy of Sciences. 1982. *Diet, Nutrition, and Cancer.* Washington, D.C.: The Academy.

National Cancer Institute and National Institute of Environmental Health Sciences. 1978. *Estimate of the Fraction of Cancer Incidence in the United States Attributable to Occupational Factors,* Washington, D.C., mimeo.

National Research Council, Committee on Biologic Effects of Atmospheric Pollutants. 1971. *Airborne Asbestos.* Springfield, Va.: NTIS.

Newberne, P.M., and Rogers, A.E. 1976. "Nutritional modulation of carcinogenesis." In *Fundamentals in Cancer Prevention, Proceedings of the 6th International Symposium of the Princess Takamatsu Research Fund,* Tokyo: the Fund, pp. 15–40.

Newman, J.A., Archer, V.E., Saccomanno, G., Kuschner, M., Auerbach, O., Grondahl, R.D., and Wilson, J.C. 1976. "Histologic types of bronchogenic carcinoma among members of copper-mining and smelting communities." *Ann. N.Y. Acad. Sci.* 271: 260–268.

Nicholson, W., and Pundsack, F. 1973. "Asbestos in the environment. Biological effects of asbestos." Lyons France: International Agency for Research on Cancer.

Norden, A. 1979. "Diet and old age." *Scand. J. Gastroenterol.* (Suppl) 14: 22–27.

Office of Technology Assessment, Congress of the United States. 1981. *Assessment of Technologies for Determining Cancer Risks from the Environment,* Washington, D.C., Superintendent of Documents.

Office of Toxic Substances, USEPA. 1974. *Preliminary Assessment of the Environmental Problem Associated with Vinyl Chloride and Polyvinyl Chloride,* Springfield, Va.: NTIS.

Office of Toxic Substances, USEPA. 1975. *Preliminary Assessment of Suspected Carcinogens in Drinking Water: Report to Congress,* Springfield, Va.: NTIS.

Ohno, Y., and Aoki, K. 1977. "Epidemiology of bladder cancer deaths in Japan." *Gann* 68 (6): 715–729.

Ottman, R., and King, M. 1980. "Family history and breast cancer risk." Paper presented at Thirteenth Annual Meeting of the Society for Epidemiologic Research, Minneapolis: June 18, 1980.

Page, G.W. 1981 "Comparison of groundwater and surface water for patterns and levels of contamination by toxic substances." *Environ. Sci. Technol.* 15: 1475–1481.

Page, R. 1973. *Economics of Involuntary Transfers.* New York: Springer-Verlag.

Page, T., Harris, R., and Epstein, S. 1976. "Drinking water and cancer mortality in Louisiana." *Science* 193: 55–57.

Peacock, P.B. 1976. "Environmental risks related to cancer." In J. W. Cullen, B.H. Fox, and R.N. Isom, eds. *Cancer: the Behavioral Dimensions.* New York: Raven Press, 85–92.

Pike, M., Gordon, R., Henderson, B., Merick, H., and SooHoo, J. 1975. "Air pollution." In J. Fraumeni, Jr., ed. *Persons at High Risk of Cancer,* New York, Academic Press, pp. 225–239.

Prindle, R. 1959. "Some considerations in the interpretation of air pollution health effects data." *J. Air. Pollut. Contr. Assoc.* 9: 12–19.

Repace, J., and Lowrey, A. 1980. "Indoor air pollution, tobacco smoke, and public health." *Science* 208: 464–472.

Rigdon, R.H., and Neal, J. 1971. "Tumors in mice induced by air particulate matter from a petrochemical industrial area." *Tex. Rep. Biol. Med.* 29 (1): 109–123.

Rogers, P. 1978. "Address." In C. Russell, ed. *Safe Drinking Water: Current and Future Problems,* Washington, D.C.: Resources for the Future, Washington, D.C. pp. 4–10.

Rohl, A., Langer, A., and Selikoff, I. 1977. "Environmental asbestos pollution related to use of quarried serpentine rock." *Science* 196: 1319–1322; rejoinder 197: 717–718.

Salg, J. 1977. "Cancer mortality rates and drinking water quality in the Ohio River Valley Basin." Ph.D. thesis, University of North Carolina at Chapel Hill, Ann Arbor, Michigan, University Microfilms.

Sawicki, E., and Westphal, D. 1961. *Symposium on the Analysis of Carcinogenic Air Pollutants,* 3 Vols., Springfield, Va.: NTIS.

Sawicki, E. 1977. "Chemical composition and potential 'genotoxic' aspects of polluted atmospheres." *IARC Sci. Publ.* 16: 127–156.

Schiffman, R., and Landau, E. 1961. "Use of indexes of air pollution potential in mortality studies." *J. Air Pollut. Cont. Assoc.* 11: 384–386.

Schneiderman, M. 1978. "Water and epidemiology," In C. Russell, ed. pp. 111–148. See Rogers (1978) for full reference.

Seidman, H. 1971. "Cancer mortality in New York City for country-of-birth, religious, and socioeconomic status." *Environ. Res.* 4 (5): 390–429.

Selikoff, I. 1968. "Asbestos." Paper presented at Symposium on Unanticipated Environmental Hazards Resulting from Technological Intrusions, Dallas, Tex., December 28, 1968.

Selikoff, I.J. 1977. "Air pollution and asbestos carcinogenesis: Investigation of possible synergism." *IARC Sci. Publ.* 16: 247–253.

Shackelford, W., and Keith, L. 1977. *Frequency of Organic Compounds Identified in Water.* Washington, D.C.: U.S. EPA, EPA-600/4-76-062, Springfield, Va.: NTIS.

Shigematsu, T., and Yamasaki, M. 1978. "Relation of occupations to the regional differences of lung cancer mortality in Fukuoka Prefecture." *Japan J. Ind. Health* 19: 182–188.

Simon, D., Yen, S., and Cole, P. 1975. "Coffee drinking and cancer of the lower urinary tract." *J. Natl. Can. Inst.* 54 (3): 587–591.

Sklar, L., and Anisman, H. 1979. "Stress and coping factors influence tumor growth." *Science* 205: 513–515.

Sorenson, A., and Lyon, L. 1979. "A case-control study of diet and cancer." Paper presented at Twelfth annual meeting, Society for Epidemiologic Research, New Haven, Conn., June 13, 1979.

Spangler, C., and DeNevers, N., for U.S. EPA. 1975. *Benzo (a) pyrene and Trace Metals in Charleston, South Carolina,* Springfield, Va.: NTIS.

Stocks, P. 1960. "On the relations between atmospheric pollution in urban and rural localities and mortality from cancer, bronchitis and pneumonia, with particular reference to 3,4 benzopyrene, beryllium, molybdenum, vanadium, and arsenic." *Brit. J. Can.* 14: 397–418.

Stocks, P., and Campbell, J. 1955. "Lung cancer death rates among non-smokers and pipe and cigarette smokers." *Brit. Med. J.* 2: 923–928.

Struba, R., and Shy, C. 1979. "Cancer and drinking water quality in North Carolina: A case-control approach utilizing prior water use exposure gradients." Paper presented at Society for Epidemiologic Research, New Haven, Conn., June 14, 1979.

Tarone, G., and Gart, J. 1975. Review of the Implications of Cancer-Causing Substances in Mississippi River Water by R. Harris, National Cancer Institute, Bethesda, Md.

Tobey, J. 1932. *Cancer: What Everyone Should Know About It.* New York: Alfred Knopf.

U.S. National Research Council. 1978. "Epidemiological studies of cancer frequency and certain organic constituents of drinking water, a review of recent literature published and unpublished." Springfield, Va, National Technical Information Service.

Vakil, D.V., and Morgan, R.W. 1973. "Etiology of breast cancer, II, Epidemiologic aspects." *Canad. Med. Assoc. J.* 109: 201–206.

Valaoras, V., MacMahon, B., Trichopoulous, D., and Polychronopoulou, A. 1969. "Lactation and reproductive histories of breast cancer patients in Greater Athens 1965–1967." *Internat. J. Can.* 4: 350.

Walker, A., and Burkitt, D. 1976. "Colon cancer: Epidemiology." *Semin. Oncol.* 3: 341–350.

Waller, R.E. 1972. "The combined effects of smoking and occupational or urban factors in relation to lung cancer." *Ann. Occup. Hyg.* 15: 67–71.

Weiss, W. 1978. "Lung cancer mortality and urban air pollution." *Am. J. Pub. Health* 68 (8): 773–775.

Wellington, D., Macdonald, E., and Wolf, P. 1979. *Cancer Mortality: Environmental and Ethnic Factors,* New York: Academic Press.

Wilkins, J., III, Reiches, N., and Kruse, C. 1979. "Organic chemical contaminants in drinking water and cancer." *Amer. J. Epidemiol.* 110: 420–448.

Wynder, E., and Hammond, E. 1962. "A study of air pollution carcinogenesis. II. Analysis of epidemiological evidence." *Cancer* 15: 79–92.

Wynder, E.L., Kmet, J., Dungal, N., and Segi, M. 1963a. "An epidemiological investigation of gastric cancer." *Cancer* 16 (11): 1461–1496.

Wynder, E.L., Onderdonk, J., and Mantel, N. 1963b. "An epidemiological investigation of cancer of the bladder." *Cancer* 16 (11): 1388–1407.

Wynder, E. 1969. "Identification of women at high risk for breast cancer." *Cancer* 24: 1235.

Yuasa, S., and MacMahon, B. 1970. "Lactation and reproductive histories of breast cancer patients in Tokyo, Japan." *Bull. WHO* 42: 195.

2 Cancer Mortality Data, Rates, Indices, and Reliability

CANCER MORTALITY DATA

Although this volume analyzes mortality data, it is pertinent to begin with a brief review of mortality data and morbidity data because the latter will be used increasingly as state incidence surveys become more widespread. Detailed discussions of both types of data are found in standard epidemiology texts (Lilienfeld and Lilienfeld, 1980; Fox and Hall, 1970).

Measures of Disease: Mortality, Incidence, and Prevalence

The number of recorded deaths (death certificates) is a measure of mortality. Computerized files of all deaths recorded as due to cancer were obtained for this study from the National Cancer Institute for the period 1950–75. The major advantage of mortality data is that it is available for almost the entire United States and almost a full 26 years. "Almost" is repeated here because the data on Hawaii and Alaska and one-half the deaths in 1972 are missing from the computerized files. This was taken into account by calculating the U.S. population base line without the two states and using only one-half of the estimated 1972 population at risk in the calculations. More specific problems were found in parts of the United States. For example, white/nonwhite distinctions were not recorded on New Jersey certificates for several years. McKay (1977) however, approximated the records by using the years immediately before and after the gap. In Virginia, many independent cities were formerly part of counties; to prepare uniform records for the state McKay reaggregated the records.

Availability is the overriding advantage of mortality data, but there are three disadvantages. First, population migration can be a substantial factor, as it was during the 26-year study period. Our record of residence is the final place of residence. The second major problem is the accuracy of the primary cause of death recorded on the certificate. The third is latency between exposure, occurrence, and death. Mortality closely reflects incidence for quickly fatal tumors (e.g., pancreas, lung), but treatment has improved for many types of cancer (e.g., leukemia, Hodgkin's disease) and in these cases there is much less association between mortality and incidence and, in turn, less association with local etiological factors.

Incidence is the number of cases occurring during a specific time period, usually a year. In comparison to mortality data, the three problems noted above are reduced with incidence data. Migration is less of a problem because a case history can be taken. Diagnosis should be more certain because tumors are usually more confined when initially diagnosed. The impact of treatment is eliminated.

If sufficient incidence data are available, then prevalence can be determined. Prevalence is usually measured as the number of people who have the disease during the time studied. Incidence and prevalence data are preferred by researchers interested in testing for associations between patterns of disease and patterns of risk.

In the absence of sufficient incidence data, mortality counts are the only option. The researcher can aggregate diseases, time periods, and places to try to minimize the problems of migration, accuracy of diagnosis, local variations in rates of cure, and latency period. These choices are discussed in the next section.

Classification of Cancer Diseases

Attempts to devise an internationally accepted system of classifying diseases date from the mid-nineteenth century. A century later, a conference held in Paris under the auspices of the World Health Organization adopted the sixth revision of the International Lists of Diseases and Causes of Death, (ICD) (World Health Organization, 1948). The format for the seventh revision made in 1955 was almost identical (World Health Organization, 1957): 3-digit codes included over 600 disease and death, about 150 injury, and almost 200 nature of injury categories. A list of 4-digit categories was also made available

An eighth revision has been used, starting in 1968 (World Health Organization, 1967). It used the same 3-digit form as the sixth and seventh revisions, but provided many additional codes for specific causes. A ninth revision appeared during the late 1970's (World Health Organization, 1977). Kurtzke (1979) has labeled ICD 9 a "regression toward a less specific and more symptom-oriented code than its predecessors."

The mortality data for 1950–75 were tabulated under the sixth, seventh and eighth revisions adapted for the United States. The National Center for Health Statistics and the National Cancer Institute used a combination of 3-digit sixth and seventh revision codes to prepare their base-line data for the period 1950–69. Rather that reclassify more than 5.3 million cancer deaths that occurred during this period, the 1970–75 data were made comparable with the 1950–69 classification. The 1970–75 data were back-fitted to the sixth and seventh revision classifications. This reclassification of 1.9 million cancer deaths was accomplished through the use of conversion manuals (Task Force for Manual of Tumor Nomenclature of Coding, 1968a,b) and the advice of Constance Percy, a member of the subcommittee for making the conversions from the seventh to the eighth and from the eighth to the seventh revision.

The result is the cancer mortality classification presented in Table 2–1, which consists of 30 male and 31 female categories, an all other category, and a total category. The basic components were age-, sex-, and race-specific mortality counts for more than 3,000 counties of the continental United States for 1950–54, 1955–59, 1960–64, 1965–69, and 1970–75 at the county scale. The mortality counts were sorted into 17 age groups: 0–4, 5–9, 10–14, 15–19, 20–24, 25–29, 30–34, 35–39, 40–44, 45–49, 50–54, 55–59, 60–64, 65–69, 70–74, 75–84, and 85$^+$. These 17 categories permitted the calculation of age-specific and age-adjusted rates. Records were available for males and females.

Finally, the counts were divided into white and nonwhite categories. Although far from the ideal, this dichotomy was the only available racial breakdown for the entire study period. The shortcomings of the white and nonwhite breakdown are discussed as necessary. At this point, it should be noted that both populations are heterogeneous. Both populations include native-born, foreign-born, and foreign stock populations.

The Advantages and Disadvantages of
Different Levels of Disease Aggregation

What are the advantages and disadvantages of using the 3-digit ICD codes, or in other words, of assuming that each cancer mortality can be traced to one of the major types of cancer listed in Table 2–1? [The alter-

Table 2-1. Cancer Mortality Causes Examined in the Study

Cancer Sites (ICD, 7th Revision)	Description
1. 140	Lips
2. 142	Salivary glands
3. 146	Nasopharynx
4. 141, 143–145, 148	Tongue; floor of mouth; other parts of mouth and mouth unspecified; oral mesopharynx; pharynx, unspecified
5. 150	Esophagus
6. 151	Stomach
7. 153	Large intestine, except rectum
8. 154	Rectum
9. 155	Biliary passages and liver (stated to be primary site)
10. 157	Pancreas
11. 160	Nose, nasal cavities, middle ear, and accessory sinuses
12. 161	Larynx
13. 162, 163	Trachea, and bronchus and lung specified as primary; and lung and bronchus, unspecified as to whether primary or secondary
14. 170	Breast
15. 171	Cervix uteri
16. 172–174	Corpus uteri; tumors of other parts of uterus including chorionepithelioma; and uterus, unspecified
17. 175	Ovary, Fallopian tube, and broad ligament
18. 177	Prostate
19. 178	Testis
20. 180	Kidney
21. 181	Bladder and other urinary organs
22. 190	Melanoma of skin
23. 191	Other skin
24. 192	Eye
25. 193	Brain and other parts of nervous system
26. 194	Thyroid gland
27. 195	Other endocrine glands
28. 196	Bone (including jaw bone)
29. 197	Connective tissue
30. 201	Hodgkin's disease
31. 200, 202, 205	Lymphosarcoma and reticulosarcoma; other forms of lymphoma (reticulosis); and mycosis fungoides
32. 203	Multiple myeloma (plasmocytoma)
33. 204	Leukemia and aleukemia
34. 147, 152, 158, 159, 164, 165, 176, 179, 198, 199	All ICD's not previously listed
35. All	All malignant neoplasms

Source: Task Force for Manual of Tumor Nomenclature of Coding, *Conversion Table Number 7B Conversion of Malignant Neoplasm, 8th Revision to Malignant Neoplasm Section, 7th Revision,* New York: American Cancer Society, 1968.

native was to aggregate more than 30 categories into more general groups: (1) buccal cavity and pharynx; (2) digestive system; (3) respiratory system; (4) female breast; (5) genital organs; (6) urinary organs; (7) central nervous system; (8) connective tissue and skin; (9) lymphatic system and hematopoietic tissue; (10) all other types; and (11) all malignant neoplasms.]

The disadvantage is the loss of information by aggregation, the advantage is increased reliability of the resulting rates. On the one hand, by classifying a cause of death as digestive cancer rather than stomach, pancreas, or liver cancer, one may lose the link to a risk factor or mix diseases that usually occur in the young with diseases that usually occur in the old. On the other hand, if the cancer is extensive and invasive, the normal landmarks are likely to be disrupted, and misclassification results. For example, the anterior border of the pancreas and the posterior wall of the stomach are adjacent. Baker (1980) suggests that the high white male pancreatic cancer mortality rate observed in St. Louis County, Minnesota may be due to incorrect classifications. Perhaps stomach cancer was the correct diagnosis, a possibility supported by the observation that the northern Minnesota region has among the highest stomach cancer mortality rates among whites. If Baker is correct, digestive cancer would have been a better classification than would pancreas or stomach cancer. Baker points to a similar problem between rectal and colon cancer. Cancer of the intestine or the digestive tract is a more reliable classification.

However, there are advantages to not combining organs, and to further disaggregating organs. For example, returning to the large intestine, Baker notes the loss of information by considering the large intestine as a homogeneous site. He argues that with respect to interaction of the intestine and its contents, there is great variation along the length of the organ. Exposure to carcinogens is different in the cecum (beginning) than in the sigmoid (last curve).

The most reasonable means of evaluating the advantages and disadvantages of a 3-digit ICD or a more general classification is to review the empirical evidence on the accuracy of death certificates. In 1940, Swartout and Webster questioned the accuracy of the death certificate as a source of information on the cause of death. Pathological examinations of 8,080 cases recorded in Los Angeles County from 1933 to 1937 revealed that 21 percent of the cases had been misclassified, but only 11 percent of the cancer cases were improperly diagnosed and/or recorded. Among the cancers, those of the respiratory system manifested the worst record: 18 percent had been misclassified.

During the next three decades, about ten papers reported on the accuracy of the death certificate. Four of these have enough cases and provide sufficient detail to be indicators of the probable difference between the 3-digit and the more general classification alternatives (Moriyama et al., 1958; Britton, 1974; Engel et al., 1980; Percy and Stanek, 1979).

The most important accuracy evaluation was presented by Percy and Stanek (1979). The study was based on 82,169 deaths drawn from 281,000 cases of the third national cancer survey conducted by the National Cancer Institute for 1969–71. Rather than the autopsies used in nearly all the other studies, the researchers began with hospital diagnoses, 90 percent of which were confirmed microscopically. When a patient died, the hospital diagnosis was compared with the underlying cause of death reported on the death certificate.

Of 82,169 deaths diagnosed as cancer in the hospital, 86 percent had been recorded as due to cancer on the death certificate, 9 percent as due to non-cancerous causes, but with cancer listed on the certificate; 5 percent had no mention of cancer on the death certificate. With respect to the 14 percent that did not list cancer as the primary cause, 63 percent were recorded as due to heart and other circulatory diseases, 11 percent as respiratory, 8 percent as digestive, and 3 percent as benign tumors. A review of these 11,447 cases at the 3-digit code level shows a variation from a low of about 10 percent misclassification for respiratory system cancers to a high of about 20 percent for brain and other central nervous system cancers. The deaths of a total of 318 persons who actually died from malignant neoplasms of the brain and other parts of the central nervous system were listed as due to another cause on the death certificate. Over one-half of these had been recorded as benign or unspecified brain tumor. Percy and Stanek speculate that some physicians do not realize that the designation "brain tumor" leads to a certificate classification "non-malignant brain tumor."

The accuracy among the more than 70 thousand cases correctly classified as cancer on the death certificate was evaluated by a detection rate and a confirmation rate. The detection rate is the number of cases listed on the death certificate compared to the number of diagnosed cases. The confirmation rate is the number of cases recorded on the death certificates that were confirmed by the diagnoses.

At one extreme, five of the specific types of cancer of particular interest to this volume had detection rates of less than 80 percent: rectum, liver, buccal cavity, larynx, and cervix. Only 54 percent of the diagnosed rectal cancer cases had been properly recorded on the death certificate. Nearly all the errors involved the designation "large intestine" instead of "rectum." The four other highest error sites ranged from 76 to 79 percent detection. The problem in each case is assigning some of the cases to an adjacent organ: larynx when it should have been buccal cavity; buccal cavity when it should have been larynx; other uterine cancer when it should have been cervix; and gallbladder and pancreas when it should have been liver.

Seven of the categories of cancer had detection rates exceeding 93 percent: esophagus, lung, female breast, prostate, brain and other central nervous system, and multiple myeloma and leukemia.

Using the above five studies, and especially the last, as the basis for comparing the 3-digit ICD with the more general classification leads to the following observations. Cancer diagnoses at the 3-digit ICD code level are about 70 percent accurate, with a range from less than 50 percent for rectal cancer to over 90 percent for cancer of the lung and the female breast, leukemia, and perhaps several others. If the more general classification, (e.g., digestive, urinary) is used, accuracy improves to about 80 percent for almost all categories. The 10 percent improvement in accuracy is appealing. However, when the author weighed this improvement against the substantial loss of information from more than 30 types down to 11 types, it was decided to use the 3-digit codes.

Two other noteworthy observations were derived from the review of accuracy of the diagnosis literature. It is interesting that cancer seems to be the most accurately reported major cause of death. Second, there is the often mentioned, but rarely measured difference between the accuracy of diagnosis and reporting in rural and urban areas. Although this difference cannot be corrected, it has been controlled to some extent by placing almost all the secondary and unspecified categories into an other cancer category. Moreover, if as noted above, the difference is a constant 10 percent, the problem does not seem to be of sufficient magnitude to invalidate the temporal comparisons between urban and rural, which are based on millions of deaths.

The Advantages and Disadvantages
of Temporal and Spatial Aggregation

As noted earlier in the chapter, misclassification of the cause of death on the certificate is not the only problem with the use of mortality data. Briefly recapitulating, the three major problems with mortality data are diagnosis and recording errors on the death certificate, population migration, and the long latency periods of most cancers. The relationship between reliability and number of mortalities is a fourth important factor influencing the reliability of rates. Aggregation of diseases is not the only solution to minimizing errors caused by these problems. Whereas aggregation of diseases is the principal method of reducing the problem of misclassification, aggregation of time periods and places are important methods used to control the problems of latency, population migration, and number of mortalities.

The latency and migration problems should be considered together because the longer the latency period the greater the effect of migration. Polissar (1980) has modeled the joint effect of the two on risk at the state, county, and place scales. Drawing upon the literature for latency period and migration assumptions, the results are expressed as a percentage of the relative excess risk of cancer incidence retained by type of geographical area. The county and state results for the 10-, 20-, and 30-year periods are reproduced here (Table 2–2).

Three factors strongly influence the results: age of the population at risk, type of area, and latency period. First, the older the population, the

Table 2–2. Relative Excess Risk of Cancer Incidence Retained in Estimated Risk by Latent Period, Anatomical Site, and Type of Geographic Area

	County			State		
Years	10	20	30	10	20	30
Site	(%)	(%)	(%)	(%)	(%)	(%)
All sites	82	68	57	90	82	72
Stomach	84	73	62	92	85	77
Colon	85	74	65	92	86	78
Rectum	84	72	62	92	85	76
Pancreas	84	73	62	92	85	77
Lung	84	71	60	91	84	75
Melanoma	77	59	46	87	75	61
Female breast	81	66	54	90	80	71
Cervix	77	59	47	87	75	63
Ovary	81	66	54	90	80	69
Prostate	85	75	66	93	87	79
Bladder	84	72	62	92	85	76
Kidney	83	70	59	91	83	74
Lymphoma	79	63	50	88	77	64
Leukemia	82	69	58	91	82	72

Source: L. Polissar, "The Effect of Migration on Comparison of Disease Rates in Geographic Studies in the United States," *Amer. J. Epidemiol.,* 111; 179 (1980).

less the probability of migration. Therefore, cancers that affect older people (e.g., prostate, rectum) have higher risk retention rates than cancers that affect the young as well as the old (e.g., leukemia, lymphoma). Second, people are less likely to cross state boundaries than county boundaries. Accordingly, county boundaries retain fewer people than do multicounty areas and state boundaries. Third, the longer the latency period, the less retention of excess risk. Polissar regards the values in Table 2–2 as minimal estimates of the reduction in relative excess risk that may occur in many cases.

Comparisons between urban and rural areas in this book probably have retention rates exceeding 90 percent because migration is minimized by the great amount of geographical aggregation. The worst case situations in this study are the suburban/central city comparisons (Chapter 6). The retention rates for these comparisons are probably between 65 and 75 percent. In order to increase the reliability of the mortality rates, both spatial and temporal aggregation have been used, sometimes in combination. Reliability will again be discussed later in the chapter when the standard error of cancer mortality rates is considered.

POPULATION AT RISK

In order to be consistent with other studies, the author followed the NCI method for estimating the population at risk.

Population at Risk, 1950–69

The data base was the *U.S. Census of Population* for 1950, 1960, and 1970. Population counts for white males and females and for nonwhite males and females for the 17 age groups were used in four formulas. The four formulas for 1950–69 are listed below:

(1) Pop. at $\text{risk}_{1950-54} = 5\,[(1960\text{ pop.} - 1950\text{ pop.})\,(.25)$
$+ (1950\text{ pop.})]$

(2) Pop. at $\text{risk}_{1955-59} = 5\,[(1960\text{ pop.} - 1950\text{ pop.})\,(.75)$
$+ (1950\text{ pop.})]$

(3) Pop. at $\text{risk}_{1960-64} = 5\,[(1970\text{ pop.} - 1960\text{ pop.})\,(.25)$
$+ (1960\text{ pop.})]$

(4) Pop. at $\text{risk}_{1965-69} = 5\,[(1970\text{ pop.} - 1960\text{ pop.})\,(.75)$
$+ (1960\text{ pop.})]$

Formula (1) adds 25 percent of the population change in, for example, the 0- to 4-year age group to the 1950 0-4 population. The result is the midyear 1952 0-4 population. The midyear 1952 estimate is multiplied by five to yield an estimate of the total 0-4 1950-54 population at risk. The second formula estimates the 1957 population, the third the 1962, and the fourth the 1967. Each of these is multiplied by five to get the five-year, age-specific population at risk for the white male and female populations. There are 17 population at risk estimates for each sex for each time period.

As an illustration, the white male population at risk in the ages 65-69 for Middlesex County, New Jersey is calculated for 1965-69. Applying formula (4), we obtain 27,419.

Middlesex County

$$\text{Pop. at risk}_{1965-69 \atop \text{age } 65-69} = 5\left[(5,580 - 5,195)\,(.75) + (5,195)\right]$$

$$= 5\left[(385)\,(.75) + (5,195)\right]$$
$$= 5\left[5,483.75\right]$$
$$= 27,419$$

Population at Risk, 1970-75

The 1970-75 population at risk was taken from estimates made for the NCI by Richard Irwin of the U.S. Census Bureau (1980). The Bureau of the Census labels the estimates as "experimental" because they have not been tested against the next census. They were prepared by the component method. The 1970 population was carried forward by age with registered births and deaths as the data. Migration, the most difficult problem, was estimated from historical data. The population estimates were also adjusted to be consistent with federal, state and county ongoing estimate programs.

Although there is a racial breakdown, Irwin cautions against the use of nonwhite data that could not be tested against available surveys. In addition, it must be noted that the Census Bureau's estimate will not be identical to numbers derived from the 1980 census.

INDICES

All the rates and indices used in this book are commonly found in the literature. Nevertheless, to make sure that the reader knows exactly what methods were used, all the formulas are indicated below and illustrated with data from the author's home county, Middlesex County, New Jersey.

Age-Specific Rate and Index

The data allowed the calculation of age-specific mortality rates for 17 age groups for five time periods. For example, the age-specific lung cancer mortality rate for the white male population of Middlesex County aged 65–69 for the years 1965–69 is 324.59.

$$\text{Age-specific rate} = \frac{\begin{array}{c}\text{Number of mortalities}\\ 1965\text{--}69\end{array}}{\begin{array}{c}\text{Population at risk}\\ 1965\text{--}69\end{array}} \times 100,000$$

$$= \frac{89}{27,419} \times 100,000 = 324.59$$

To aid in comparisons, an age-specific rate index was calculated. The county age-specific rate is divided by the national age-specific rate for the same population. The resulting index allows better visual comparison of rates for different cohorts. For example, the age-specific lung cancer mortality rates for the white male population aged 45–54 and 75+ of Middlesex County are 52.5 and 337.8, respectively. These numbers are far more meaningful in the form of the age-specific rate index: 107 (7 percent higher than the nation) for the 45–54 cohort, 187 (87 percent higher than the nation) for the 75+ population. Examples of other comparisons facilitated by this index are between urban and rural and central city and suburban.

Age-Adjusted Rate and Index

The age-adjusted rate is the most commonly used summary index of incidence and mortality. The calculation of the age-adjusted, lung cancer mortality rate for white males in Middlesex County, New Jersey will be used as an illustration (Table 2–3).

The number of mortalities (column 1) is divided by the population at risk (column 2) to yield the age-specific rate (column 3). The age-specific rates are multiplied by the standard vector (column 4). In this case, the standard vector is the 1960 population of the United States. The results

Table 2–3. Illustrative Calculation of an Age-Adjusted Rate

Age Groups	(1) Trachea, Lung, Bronchus Deaths, 1965–69	(2) Population at Risk, 1965–69	(3) Age-Specific Rate, 1965–69 per 100,000	(4) Standard Population Vector, 1960	(5) Age-Adjusted Rate, 1965–69 per 100,000
1. 0–4	0	126,790	0	.11332	0.0
2. 5–9	0	140,671	0	.10423	0.0
3. 10–14	0	136,878	0	.09354	0.0
4. 15–19	0	114,248	0	.07372	0.0
5. 20–24	0	84,033	0	.06023	0.0
6. 25–29	1	86,673	1.15	.06061	0.0697
7. 30–34	1	83,584	1.20	.06663	0.0799
8. 35–39	9	88,368	10.18	.06960	0.7085
9. 40–44	14	94,483	14.82	.06469	0.9587
10. 45–49	36	89,139	40.39	.06067	2.4505
11. 50–54	58	75,739	76.58	.05357	4.1024
12. 55–59	93	57,583	161.51	.04701	7.5925
13. 60–64	88	40,103	219.43	.03983	8.7399
14. 65–69	89	27,419	324.59	.03490	11.3282
15. 70–74	91	19,630	463.58	.02643	12.2524
16. 75–84	101	19,694	512.85	.02584	13.2520
17. 85+	10	3,649	274.05	.00518	1.4196
Total	591	1,288,684	—		62.9543

(column 5) are added to yield the age-adjusted rate 62.95. By comparison, the 1950–54 white male lung cancer rate is 39.68.

An age-adjusted rate index was obtained by dividing the county rate by the national age-adjusted rate. A different impression of the 1950–54 and 1965–69 rates (39.68 and 62.95) follows the division: 156 in 1950–54 and 126 in 1965–69. Thus, although Middlesex's white male lung cancer rate rose dramatically during the two decades, the nation's rate rose even more.

Proportional Mortality Rate

In order to evaluate the changing importance of each cause over time, the number of cancer mortalities of each type is divided by the total number of cancer-related deaths. For example, during 1950–54, lung cancer accounted for 9 percent of mortalities among white males at least 75 years old. Stomach cancer was a more important cause during 1950–54: 15%. By 1965–69, lung cancer rose to almost 23 percent, whereas stomach cancer fell to less than 10 percent.

Time Change Index

This index measures the relative change in cancer mortality over time. It is obtained by dividing one rate, for example, the 1965–69 age-adjusted rate by another, for example, the 1950–54 rate, and multiplying by 100. For example, assume the index for white male lung cancer is 159, whereas for stomach cancer it is 50. Lung cancer rates increased almost 60 percent, whereas the stomach cancer rate in 1965–69 is one-half of its 1950–54 total.

Reliability: Standard Error and the Ecological Fallacy

The goals of this research require that three questions of reliability be answered together. First, how well does a single rate represent a region? Second, how confident may we be with a finding that the rates of two places are different? Third, even if the statistical reliability of the rates is known, how sure can we be that the urban cancer mortality link is really being tested by the data configurations?

With respect to the first question, a representative mortality rate is one that is not seriously changed by unrepresentative, random events. The events that influence the rates include births, deaths, migration, and diagnoses. Since the analyst cannot change these and other determinants of cancer rates after the fact, increasing the population at risk, by combining diseases, places, and time periods, can reduce random factors that could cause unrepresentative mortality rates. Thus, we should have far more confidence in the rates of a region with 100,000 deaths than a region with 10,000 deaths. The added confidence in the rates of the more populous regions is reflected in the fact that they manifest smaller standard errors than the less populated regions.

Using Chiang's (1961) method for calculating standard errors of cancer rates, the NCI (Mason et al., 1975) devised a method of determining precisely how confident one can be that the rate of a region A really differs from the rate of a region B. The method can be summarized as follows: Region A's rate is significantly lower (with 95 percent confidence) than Region B's rate if A's rate plus 1.96 times the standard error of A's rate is less than B's rate minus 1.96 times the standard error of B's rate. This method is used here.

Although the confidence limits and significance rates have been calculated, they are not always presented in this book. They are presented when a specific hypothesis is being tested or when a county rate is being compared to a national or another regional rate. Confidence limits and rates of significance are not presented for the many tables that compare major urban aggregates (e.g., all central cities, all suburbs, all rural areas). This decision was made for three reasons. First, the tables are sizable even without the added confidence limit and significance rate data. Indeed, in order to increase the utility of the data in the tables, they were converted to graphics and the tables were placed in the appendices. Second, most of the aggregates have so many people that the standard errors for the major diseases are usually less than 1 percent of the rate. Third, other authors working with aggregate data (Lilienfeld et al., 1972; Mason et al., 1975) have followed the same practice.

The third question about reliability arises from generalizing about individuals from aggregate data (Blalock, 1964; Alker, 1969; and Clark and Avery, 1976). When comparing a region that is 95 percent urban with one that is 5 percent urban, it is safe to assume that urban people dominate one region and rural people dominate the other. But, not all comparisons are made between places at opposite ends of the urban/rural spectrum. If the two regions were two-thirds and one-third urban instead of 95 and 5 percent urban, it is possible that a difference in cancer mortality between the two regions could be mistakeningly attributed to urban populations when it was actually due to higher rates among the rural populations in one region than in the other. However, the results of the research described in the following chapters are so strong and consistent that it is doubtful that ecological fallacy threatens the spatial convergence trend.

REFERENCES

Alker, R., Jr. 1969. "A typology of ecological fallacies." In *Ecological Analysis in the Social Sciences* M. Dogan and S. Rokkan, eds. Cambridge, Mass.: MIT Press, pp. 69–86.

Baker, A. 1980. "Pitfalls in evaluation of cancer data for geographic analysis." Paper presented at 1980 meetings of the Association of American Geographers, Louisville, Kentucky.

Blalock, H. 1964. *Causal Inferences in Nonexperimental Research*. Chapel Hill: University of North Carolina Press.

Britton, M. 1974. "Diagnostic errors discovered at autopsy." *Acta med. scand.* 196: 203–210.

Chiang, C. 1961. "Standard error of the age-adjusted death rate." *Vital Statistics, Selected Reports* 47: 275–285.

Clark, W., and Avery, K. 1976. "The effect of data aggregation in statistical analyses." *Geograph. Anal.* 8: 428–438.

Engel, L., Strauchen, J., Chiazze, L. Jr., Heid, M. 1980. "Accuracy of death certification in an autopsied population with specific attention to malignant neoplasm and vascular diseases." *Amer. J. Epidemiol.* 111: 99–112.

Fox, J., and Hall, C. 1970. *Epidemiology: Man and Disease.* New York: Macmillan.

Kurtzke, J. 1979. "ICD 9: A regression." *Amer. J. Epidemiol.* 109: 383–393.

Lilienfeld, A., and Lilienfeld, D. 1980. *Foundations of Epidemiology,* 2nd Ed. New York: Oxford University Press.

Lilienfeld, A., Levin, M., and Kessler, I. 1972. *Cancer in the United States.* Cambridge, Mass.: Harvard University Press.

Mason, T., McKay, F., Hoover, R., Blot, W., and Fraumeni, J. Jr. 1975. *Atlas of Cancer Mortality for U.S. Counties 1950–1969,* Washington, D.C.: U.S. Department of HEW, NIHDHEW Pub. No. (NIH) 75, 780.

McKay, F. 1982. personal conversations, 1977–82.

Moriyama, I., Baum, W., Haenszel, W., and Mattison, B. 1958. "Inquiry into diagnostic evidence supporting medical certifications of death." *Amer. J. Pub. Health* 48: 1376–1387.

Percy, C., and Stanek, E. III 1979. Accuracy of cancer death certificates and its effect on cancer mortality statistics. Paper presented at 107th Annual Meeting of the American Public Health Association, Statistics Section, November 7, 1979.

Polissar, L. 1980. "The effect of migration on comparison of disease rates in geographic studies in the United States." *Amer. J. Epidemiol.* 111: 175–182.

Swartout, H., and Webster, R. 1940. "To what degree are mortality statistics dependable?" *Amer. J. Pub. Health* 30: 811–815.

Task Force for Manual of Tumor Nomenclature and Coding 1968a. Conversion Table Number 7A, Conversion of Malignant Neoplasm Section, 7th Revision of *International Classification of Diseases* to Malignant Neoplasm Section, 8th Revision of *International Classification of Diseases, Adapted,* New York, American Cancer Society.

—— 1968b. Conversion Table Number 7B, Conversion of Malignant Neoplasm Section, 8th Revision of *International Classification of Diseases, Adapted* to Malignant Neoplasm Section, 7th Revision of *International Classification of Diseases,* New York, N.Y.: American Cancer Society.

U.S. Bureau of The Census 1980. *Current Population Reports,* Series P-23, No. 103, Methodology for Experimental Estimates of the Population of Counties, by Age and Sex: July 1, 1975, Washington, D.C.: U.S. Government Printing Office.

World Health Organization 1948. *Manual of the International Statistical Classification of Diseases, Injuries and Causes of Death,* Vol. 1, Geneva: WHO.

—— 1957. *Manual of the International Statistical Classification of Diseases, Injuries and Causes of Death, Seventh Revision,* Vol. 1, Geneva: WHO.

—— 1967. *Manual of the International Statistical Classification of Diseases, Injuries and Causes of Death, Eighth Revision, Vol. 1,* Geneva: WHO.

—— 1977. *Manual of the International Statistical Classification of Diseases, Injuries and Causes of Death, Ninth Revision, Vol. 1,* Geneva: WHO.

3 A National and International Base Line

The purpose of this chapter is to provide a base line of cancer patterns in the United States during the 1950–75 study period. This has two components. The first is cancer mortality data from the continental United States for the period 1950–75. The second is a comparison of cancer mortality rates in the United States and other countries that indicates the extent to which trends reported for the United States are common to other countries.

Many excellent studies have given a detailed picture of cancer mortality and cancer incidence in the United States (for example, Burbank, 1971; Lilienfeld, Levin, and Kessler, 1972; Devesa and Silverman, 1978; McKay, Hanson, and Miller, 1982; and Pollack and Horm, 1980). In this chapter the most important of these findings will be summarized as they relate to urban areas. Age-specific rates will be given only for those diseases that particularly elucidate the differences between urban and other areas found during the study period.

Overall, this chapter seeks the answers to four questions:

1. Which types of cancer have become more important and which less important during 1950–75?
2. Are cancer mortality trends in the United States different for white males, white females, nonwhite males, and nonwhite females?
3. In which age groups are the important cancer mortality trends in the United States occurring?
4. How do cancer mortality rates and trends in the United States compare with other nations?

Although not directly germane to the relationship between urbanization and cancer mortality trends in the United States during 1950–75, the following two questions, of importance to many readers, are addressed in Appendix 3.

1. How similar are cancer mortality trends in the United States before and during the study period?
2. How similar are cancer mortality and cancer incidence trends in the United States?

CANCER MORTALITY IN THE UNITED STATES, 1950–75

Pronounced shifts have occurred in the relative importance of different types of cancer. These changes will be described in two parts. First, proportional mortality (not age adjusted) and age-adjusted mortality rate data are presented to demonstrate the changing importance of 34 types of cancer among the white male, white female, nonwhite male, and nonwhite female populations. Second, age-specific data for four selected types of cancer are utilized as a base line to compare places with different levels of urbanization.

Lilienfeld and Lilienfeld (1980) outline the possible reasons for the changing cancer trends (Table 3–1). Although some are reviewed in this chapter, precise description of the trends, not explaining them, is the goal of this chapter. See Doll and Peto (1981), Smith (1980), and Pollack and

Table 3–1. Outline of Possible Reasons for Changes in Mortality Trends of Disease

A. Artifactual
   1. Errors in the numerator due to
      (a) changes in the recognition of disease;
      (b) changes in rules and procedures for classification of causes of death;
      (c) changes in the classification code of causes of death; and
      (d) changes in accuracy of reporting age at death
   2. Errors in the denominator due to
      (a) errors in the enumeration of the population
B. Real
   1. Changes in age distribution of the population;
   2. Changes in survivorship; and
   3. Changes in incidence of disease: the result of
      (a) genetic factors and
      (b) environmental factors

Source: Lilienfeld and Lilienfeld, 1980, p. 86.

Horm (1980) for different interpretations of the trends, and Appendix 3 for an analysis of incidence and mortality trends.

Age-Adjusted Rates: Changes Among the Total Population

With respect to mortality, seven of the 34 types of cancer have become much more important, five slightly more important, eight far less important, and eight slightly less important during the study period. The remaining six types are inconsistent across the four populations when they are categorized by changing importance. These observations will be elucidated, with reference to Figures 3–1 through 3–5 and Tables A1–1 through A1–8.

SEVEN TYPES OF CANCER THAT HAVE BECOME
MUCH MORE IMPORTANT

Lung cancer is now the most important cause of cancer-related mortality. During 1950–54, lung cancer accounted for almost 16 percent of white male cancer mortalities not too far ahead of stomach cancer, which caused about three-quarters as many. By 1970–75, mortality due to lung cancer in white males had almost doubled, to more than 30 percent, accounting for more than three times as many as cancer of the large intestine, the second leading cause of cancer mortality among white males.

The marked increase in lung cancer is even stronger among the other three populations than among white males (Figure 3–1). Lung cancer was the second most important cause of death among nonwhite males during 1950–54 (cancer of the stomach was first). By 1955–59, it had become the leading cause and by 1970–75 it accounted for nearly 30 percent of nonwhite male cancer deaths, more than twice as many as prostate cancer, the second leading cause.

Whereas male lung cancer rates have been about five times as high as female rates, the most obvious relative increases in lung cancer during the study period have been among females. During 1950–54, lung cancer was the eleventh leading cause of cancer deaths among white females and the tenth among nonwhite females. Breast cancer was responsible for more than five times as many cancer deaths as lung cancer among females. One decade later, lung cancer had risen to seventh place among white females and to eighth among nonwhite females. The great magnitude of the increase did not, however, become apparent until the mid-1960's and especially the 1970's. During 1970–75, lung cancer had become the third leading cause of cancer-related deaths among white and

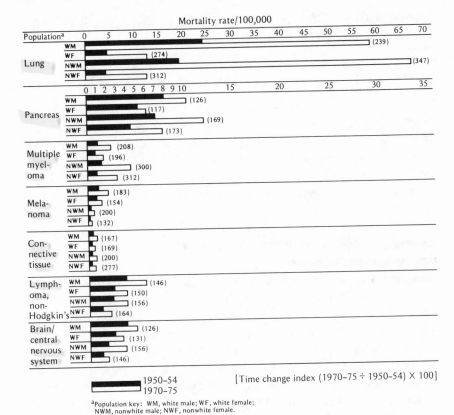

Figure 3-1. Change in age-adjusted cancer mortality of seven types that have become far more important, 1950-54 to 1970-75

nonwhite females. Although the relative increase in importance has been partially due to decreasing rates of digestive organ and especially genital organ cancers among females, the increase in the relative importance of the disease is primarily due to the tripling of the age-adjusted rates during the study period. The magnitude of the recent acceleration of the rate increase implies that lung cancer could potentially become the leading cause of cancer mortality among females, although it was responsible for only about one-half of the mortalities of breast cancer during 1970-75.

The combination of a high absolute cancer mortality rate and a high rate of increase of the rate for all four populations is unique to lung cancer, although six other types have sufficiently pronounced increases to be particularly noteworthy. Cancer of the *pancreas* is the most important cause among these six (Figure 3-1). During 1950-54, it ranked sixth among white males, tenth among white females, seventh among nonwhite males, and ninth among nonwhite females as a cause of cancer mortality. By 1970–75, it increased to the fourth or the fifth leading cause among the four subpopulations. Age-adjusted rates increased about 20 percent among the white population compared to almost 70 percent for non-whites. Cutler and DeVesa (1973) hypothesize that better reporting accounts for much of the increase of nonwhite rates.

The remaining five types were not among the ten leading causes during 1950–54. Nevertheless, each type recorded increases that should not be ignored. The largest relative increase was recorded by *multiple myeloma:* about double for whites and triple for nonwhites (Figure 3–1). As a result of this marked increase, the disease increased from an average of the twentieth most important among the four populations to an average of the fourteenth most important cause of cancer mortality.

Melanoma followed a similar path (Figure 3–1). Age-adjusted rates substantially increased for all groups: from one-third for nonwhite females to double for nonwhite males. As a result, melanoma increased in average importance from about the twenty-fourth to about the twentieth.

Among the four populations, cancer of the *connective tissue* was number twenty-second, twenty-third, or twenty-fourth in number of cancer-related deaths during 1950–54. Two decades later, due to increases ranging between 67 percent for white males to 177 percent for nonwhite females, this disease had increased in importance to between nineteenth and twenty-second place.

The sixth of the seven types of cancer that manifested substantial increases among all four subpopulations was *non-Hodgkin's lymphoma.* Age-adjusted rates consistently increased about 50 percent among the four populations (Figure 3–1). The importance of the disease as a cause of mortality increased from an average of thirteenth to an average of twelfth.

The last of the seven types characterized by substantial increases in age-adjusted cancer mortality rates during the study period is *cancer of the brain and central nervous system.* Compared to the above six types, the increases were relatively modest. The range was 26 percent for white males to 56 percent for nonwhite males.

Overall, seven types of cancer manifested substantial increases in cancer mortality among each of the four populations. During 1950–54, these seven types accounted for almost 28 percent of white male cancer-related deaths. Two decades later, these seven types had increased to almost 45 percent of white male cancer-related mortalities. The increase among nonwhite males was similar to white males: from 23 percent of all cancer deaths during the early 1950's to 40 percent two decades later. Similar changes were observed for these seven among females. The seven accounted for about 13 percent of white and 9 percent of nonwhite female mortalities during 1950–54. Two decades later, these seven increased to 24 percent for white females and 21 percent for nonwhite females. Lung cancer does account for the largest proportion of the increase among each subpopulation. However, the other six types should not be neglected because of preoccupation with lung cancer.

FIVE TYPES OF CANCER THAT HAVE BECOME
SLIGHTLY TO MODERATELY MORE IMPORTANT

Cancer of the larynx, nasopharynx, kidney, female breast, and ovary demonstrated smaller increases than the seven types of cancer discussed above. The most important of these are cancers of the female breast and ovary (Figure 3–2). *Female breast cancer* was the most important cause of female cancer-related deaths during 1950–75. During the two and one-half–decade study period, female cancer mortalities due to breast cancer increased from 18.5 to 20 percent for whites and from 15.1 to 17.2 percent for nonwhites. These small relative increases paralleled small increases in the age-adjusted rates: 2 percent for whites and 15 percent for nonwhites. Slight increases also characterized *ovarian cancer*. During

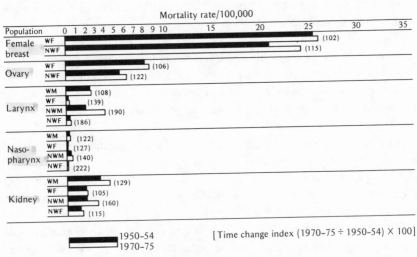

Figure 3–2. Change in age-adjusted cancer mortality of five types that have become slightly to moderately more important, 1950–54 to 1970–75

1950–54, it ranked as the sixth most important cause of cancer mortalities among females. The importance of ovarian cancer has not dramatically shifted because the age-adjusted mortality rates have only slightly increased: 6 percent for white females and 22 percent for nonwhite females.

Although female breast and ovarian cancer manifested slight, but consistent increases across the subpopulations, cancers of the *larynx, nasopharynx,* and *kidney* are characterized by substantial increases that vary by subpopulation (Figure 3–2). The first two account for about 1.8 percent of the male and 0.4 percent of female cancer-related mortalities. Age-adjusted rates of larynx and nasopharynx cancer have increased, considerably more so among nonwhites (approximately doubled) than among whites (increased about 25 percent). Cancer of the kidney accounts for about 1.5 percent of total cancer-related mortalities. Age-adjusted rates among males have been increasing at a relatively higher rate (an average of 45 percent) than the rate among females (an average of 10 percent).

EIGHT TYPES OF CANCER THAT HAVE BECOME
MUCH LESS IMPORTANT

Five of the eight types of cancer that have manifested substantial decreases in age-adjusted rates had been among the most important causes of cancer-related mortalities. During 1950–54, *stomach* cancer was the leading cause of cancer death among nonwhite males, the second among white males, the third among white females, and the fourth among nonwhite females. Stomach cancer remains a leading cause of death, but it is relatively less important today. During 1950–75, it dropped from the first to the third most important cause among nonwhite males, from the second to the fourth among white males, from the third to the seventh among white females, and from the fourth to the sixth among nonwhite females as the result of a precipitous decrease of between 40 and 60 percent in the age-adjusted rates (Figure 3–3). Most researchers hypothesize that this decrease is due to changing diets and improved preservation and handling of food.

Like stomach cancer, two other parts of the digestive system, the *rectum* and *liver,* have manifested pronounced decreases in age-adjusted rates among all four populations. Both were among the nine most important causes of cancer mortality among all four populations during 1950–54. Two decades later, both had dropped an average of four places and were usually no longer among the top ten causes. Their age-adjusted rates decreased an average of about 40 percent.

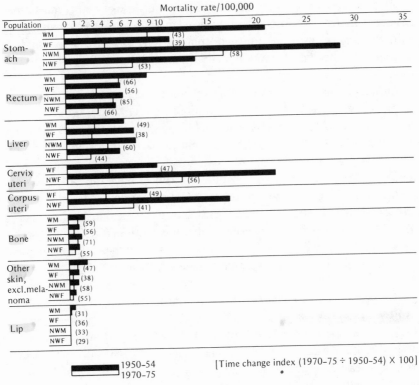

Figure 3–3. Change in age-adjusted cancer mortality of eight types that have become far less important, 1950–54 to 1970–75

Overall during the early 1950's, four parts of the digestive system were among the ten most important sites of cancer. Three of these, stomach, rectum, and liver, have exhibited marked decreases in age-adjusted rates. Whereas these three were responsible for two out of ten cancer-related mortalities during 1950–54, two decades later one out of ten was due to these three types.

Cervix uteri and *corpus uteri* cancers have drastically decreased as causes of female mortality (Figure 3–3). During 1950–54, they were responsible for 13 percent of white female and 28 percent of nonwhite female cancer mortalities. Two decades later, following decreases in age-adjusted rates of about 50 percent, the percentage of cancer deaths caused by these diseases dropped to 7 percent for white females and 14 percent for nonwhite females. Among white females, cancer of the cervix uteri dropped from the fourth most important cause to the eighth, and cancer of the corpus uteri from the fifth to the ninth. The relative drop was even more marked among nonwhite females: cancer of the cervix uteri from the first most important cause in 1950–54 to the fourth during 1970–75; and cancer of the corpus uteri from the third to the seventh.

The remaining three types of cancer, *bone, other skin* (excluding melanoma), and *lip,* are minor causes of cancer mortality and have become far less important (Figure 3–3). All had an age-adjusted rate less than 2/100,000 during 1950–54. By 1970–75, none had a rate exceeding 1/100,000. The most pronounced decreases were for cancer of the lip, a disease that primarily affects white males. Mortality rates decreased by about 70 percent among each population. Decreases for bone and other skin cancers were less than for lip cancer (between 30 and 60 percent).

In summary, the eight types of cancer reviewed in this section have manifested substantial decreases in age-adjusted mortality rates for each population during the study period. Among nonwhite females, the eight types accounted for 46 percent of cancer-related deaths during 1950–54, compared to only 24 percent during the early 1970's. The comparative figures for white females are 32 percent to 17 percent. The white and nonwhite male decrease in proportional mortality for the six types has been from about 25 percent to about 11 percent. Overall, the eight types were only about one-half as important as causes of cancer-related deaths during 1970–75 as they were during 1950–54.

EIGHT TYPES OF CANCER THAT HAVE BECOME
SLIGHTLY TO MODERATELY LESS IMPORTANT

Hodgkin's disease and seven minor types of cancer (*thyroid, nose, eye, testis, male breast, other endocrine,* and *salivary glands*) slightly to moderately decreased in importance between 1950–54 and 1970–75 (Figure 3–4). In almost every case (five of eight), the smallest decrease in the age-adjusted rates for these eight types was observed for nonwhite males. The average decrease during the study period for these eight types among nonwhite males was 10 percent, with a range from a 7 percent increase (salivary glands) to a 20 percent decrease (nose). If nonwhite males are at one end of the spectrum with respect to these eight diseases, white females are at the other end. The average decrease among white females was 27 percent with a range from 10 percent (other endocrine) to 45 percent (nose).

White male and nonwhite female rates of decrease were between the above two. The average decrease for white males was 19 percent, for nonwhite females 17 percent.

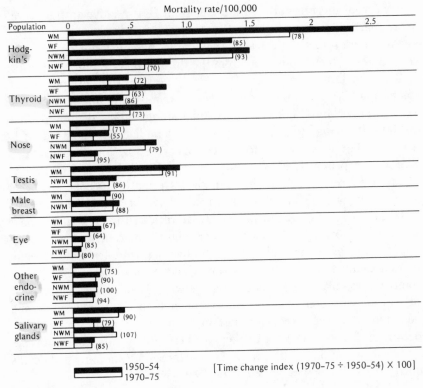

Figure 3–4. Change in age-adjusted cancer mortality of eight types that have become slightly to moderately less important, 1950–54 to 1970–75

The slight to moderate decreases for these eight types of cancer is most succinctly summarized by the observation that they accounted for about 2.7 percent of cancer-related mortalities during 1950–54 compared to 1.8 percent two decades later.

SIX TYPES OF CANCER THAT HAVE
EXHIBITED INCONSISTENT PATTERNS OF CHANGE

In contrast to the 28 types that tended to be consistent across the four populations, six important causes of cancer-related mortalities varied strongly in rates of change in at least one of the four populations (Figure 3–5). For three of six, *cancer of the esophagus* and of the *prostate* and *leukemia*, whites and nonwhites manifested different rates of change. Cancer of the esophagus increased more than 70 percent among non-

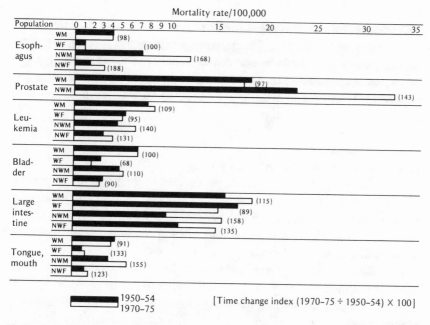

Figure 3–5. Change in age-adjusted cancer mortality of six types that have exhibited inconsistent patterns of change, 1950–54 to 1970–75

whites, while remaining stable among whites. Similarly, cancer of the prostate as a cause of death increased more than 40 percent among non-white males, while remaining stable among white males. Similarly, although leukemia rates increased about 35 percent among nonwhites, they remained relatively stable among whites.

With respect to the other three types of cancer (of the *bladder, large intestine,* and *tongue/mouth*), three of the four populations have manifested similar trends. White females and to a lesser extent nonwhite females are the deviant populations for two of the four: cancer of the large intestine and bladder. Although the white female rates slightly to moderately decreased, the white male and nonwhite male populations have manifested stable to moderately increasing rates. Devesa and Silverman (1978) suggest that the male up/female down dichotomy for bladder cancer is due to male occupational exposure to carcinogens and to excessive cigarette smoking. White females and nonwhites exhibited moderate increases for the tongue/mouth category, whereas the white male population manifested a slight decrease.

SUMMARY
Twenty-eight of the 34 types of cancer (breast cancer was counted as two types) exhibited relatively similar trends among the four populations. Seven became far more important, five slightly to moderately more important, eight far less important, and eight slightly to moderately less important. The remaining six types exhibited an obvious deviation in at least one of the four populations.

TESTING FOR PARALLEL CHANGE AND CONVERGENCE
OR DIVERGENCE AMONG THE POPULATIONS

The data presented in the previous section led to the preliminary conclusions that the four subpopulations are following similar patterns of change, but that divergence rather than convergence has occurred between them. This section examines these preliminary conclusions in detail, using rank correlations and graphic aids.

Bivariate correlations were made using the rates of change of each disease (including all other types as a category) between 1950–54 and 1970–75 as data. Rank correlation was used rather than parametric correlation methods for two reasons (see Appendix 2 for a detailed discussion). On the one hand, if each disease were made equally important, then a few low rate diseases would overly influence the results. On the other hand, if weighted parametric regression methods were used to control the first problem, then lung cancer, the disease with the highest rate of increase and the highest initial rate, would dominate the results. This double dilemma led to the adoption of the more conservative rank order statistic. The disease types were ranked from highest increase during the study period (lung) to highest decrease (lip) for each population.

The results provide two pieces of information: (1) the higher the correlation, the greater the similarity between the two subpopulations in rates of change; (2) those diseases that do not consistently exhibit a similar rank might require special attention. All six combinations of the four populations were tested.

The results imply a strong degree of parallel change among the four populations (Table 3–2). All the correlations are significant at the .01 level, with cancer of the lung and multiple myeloma ranked first and second and cancer of the lip ranked last for each population.

Although correlations are extremely high, some marked differences in rank among the four subpopulations were observed. Using a difference in rank of five or more between any two populations as an indicator of strong deviation, 13 types, including five of the six types previously qual-

Table 3–2. Rank Correlation of Change in Cancer Mortality Rates between the Four Subpopulations, United States, 1950–54 to 1970–75

Subpopulations	Spearman Rank Correlation	Significance	Strong Deviants		
1. White male vs. nonwhite male	.94	.01	Esophagus Larynx	Nasopharynx Lymphoma	
2. White male vs. white female	.92	.01	Tongue/mouth Larynx	All other Large intestine	Bladder Kidney
3. White female vs. nonwhite male	.92	.01	Large intestine Tongue/mouth	Lymphoma Nasopharynx	Esophagus All other
4. White female vs. nonwhite female	.91	.01	Esophagus Lips Large intestine	Melanoma Tongue/mouth Nasopharynx	Hodgkin's disease
5. Nonwhite male vs. nonwhite female	.90	.01	Nasopharynx Melanoma	Lips Kidney	All other
6. White male vs. nonwhite female	.87	.01	Esophagus Kidney Melanoma	All other Larynx Nasopharynx	Nose

itatively classified as inconsistent and eight other types, were identified as deviants. Cancer of the nasopharynx is a significant deviant in five of the six correlations, followed by cancer of the esophagus and all other types of cancer, which appear four times, and cancer of the larynx, tongue/ mouth, large intestine, kidney, and skin (melanoma), which are strong deviants three times. Lip cancer and lymphoma are deviants twice and bladder cancer, Hodgkin's disease, and nose cancer once.

Overall, the Spearman rank correlations show that the rates of change of the vast majority of types of cancer have been quite consistent across the four populations.

The strong consistency trend is paralleled by a clear trend toward divergence among several of the populations. The marked relative increase of nonwhite male and decrease of white female cancer mortality rates is apparent within every major cancer group (Figures 3–6 through 3–13). The *all types of cancer* mortality time series graph (Figure 3–6) shows that males and females have diverged and that nonwhite males have manifested marked and growing divergence from the other three subpopulations. During 1950–54, white males had the highest all types of cancer mortality rate: 7 percent higher than nonwhite males; 15 percent higher than nonwhite females; and 18 percent higher than white females. Two decades later, the gap between white males and both female populations had more than doubled.

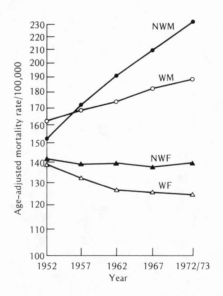

Figure 3–6. Trends in all types cancer mortality, 1950–75

Although noteworthy, the divergence between white males and the two female populations is overshadowed by the clearly increasing gap between nonwhite males and the other populations. During 1955–59, the age-adjusted rate of the nonwhite male population exceeded that of the white male population. During 1970–75, the differences between the nonwhite male population and the white male, nonwhite female and white female age-adjusted rates were 22, 65, and 86 percent, respectively.

Nonwhite, especially nonwhite male, cancer mortality rates were undoubtedly seriously undercounted during the 1950's. However, the most substantial increase in the gap between the other three populations and nonwhite males occurred during the late 1960's and early 1970's, which implies that improved data gathering could not fully explain the marked divergence between nonwhite males and the other three populations. During a 26-year period, while the total cancer mortality rate of white females was decreasing 10 percent, the nonwhite female rate was decreasing 1 percent; and while the white male rate was increasing 16 percent, a set of risk factors must have greatly affected the nonwhite male population in order to account for an increase in the nonwhite male total population, age-adjusted rate of more than 50 percent.

The major *respiratory* cancer types (of the lung, larynx, and nose) exemplify parallel trends among all four populations as well as divergence for nonwhite males (Figure 3–7). The rates of all populations more than doubled during the study period. Only the nonwhite male rate more than tripled. The white male rate almost doubled between 1950–54 and 1965–69. Yet by 1965–69, it fell below the nonwhite male rate. By 1970–75, the white male rate was almost 20 percent less than the nonwhite male respiratory cancer mortality rate.

Although the gap between nonwhite males and the other three populations has increased, the gap between the other three populations has narrowed. During 1950–54, the white male respiratory cancer mortality rate was 5.3 times the white female and 6.0 times the nonwhite female rate. Two decades later, the difference was still pronounced, but had decreased to 4.6 times.

The major cancers of the *digestive* system (esophagus, stomach, large intestine, rectum, liver, and pancreas) convey the same message as those of the respiratory system, despite the fact that the rate for this system has manifested marked decreases, whereas the rates for the respiratory system have rapidly increased (Figure 3–8). The nonwhite male rate has remained stable, and the white female rate has markedly decreased. This pattern is remarkably consistent across the six types of cancer. The nonwhite male population exhibited the highest increase or lowest decrease

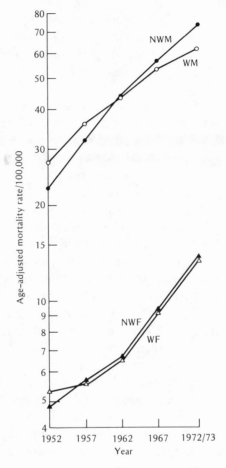

Figure 3–7. Trends in respiratory cancer mortality, 1950–75

for five of the six types, the white female the smallest increase or most pronounced decrease for five of the six types.

The widening of the gap is especially pronounced for *cancer of the esophagus*. During 1950–54, the nonwhite male rate was 1.7 times the white male rate, 4 times the nonwhite female rate, and almost 7 times the white female rate. During the following two decades, the nonwhite male rate increased 68 percent, and the white population rates stabilized. By 1970–75, the age-adjusted mortality rate for cancer of the esophagus among nonwhite males was three times the white male rate, four times the nonwhite female rate, and eleven times the white female rate. The widening of the gap between nonwhites and whites leads to the obser-

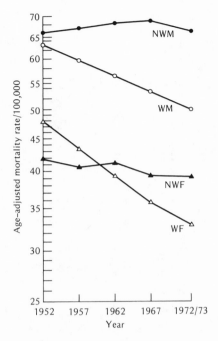

Figure 3–8. Trends in digestive system cancer mortality, 1950–75

vation that esophageal cancer has become the most race-differentiated type of cancer.

Although cancer of the esophagus most clearly exemplifies divergence of nonwhite males from the other three populations, the same pattern has been established for the other five major digestive system cancer types. The white and nonwhite male total digestive system cancer mortality rates were virtually identical during the beginning of the study period. By 1970–75 the nonwhite male rate was 32 percent higher than the white male rate.

The major *genital cancers* (prostate, testis, cervix uteri, corpus uteri, and ovary) closely replicate the digestive system cancer pattern (Figure 3–9). However, divergence for nonwhite males is even more marked than for digestive system cancers. Although nonwhite male rates increased more than 40 percent, the rates of the other three populations decreased. This result is due to cancer of the prostate rates, which increased 43 percent among nonwhite males, while slightly decreasing among white males. The gap between nonwhite males and white males increased from 26 to 86 percent during the study period.

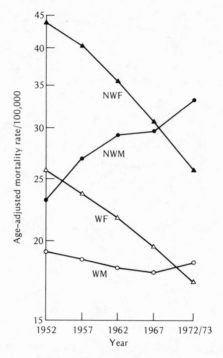

Figure 3–9. Trends in genital cancer mortality, 1950–75

The gap between white female and nonwhite female all types of cancer mortality has increased from 2 percent during 1950–54 to 12 percent during 1970–75. *Female breast cancer* was a major contributing factor because the difference between white female and nonwhite female rates decreased from 22 percent during 1950–54 to 8 percent during 1970–75 (Figure 3–10).

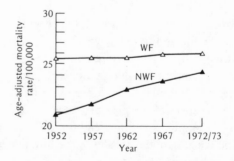

Figure 3–10. Trends in female breast cancer mortality, 1950–75

Urinary tract (bladder, kidney) cancers repeat the divergence pattern. Nonwhite male rates showed the largest increase, white male rates increased slightly, and female rates decreased, white female more than nonwhite female (Figure 3–11). The gap between males and females increased, and the gap between white males and nonwhite males decreased.

The *lymphatic* and *hematopoietic system cancers* (leukemia, lymphoma, Hodgkin's disease, and multiple myeloma) illustrate the by now expected patterns. The nonwhite male rate increased most rapidly (61 percent), the white female least rapidly (17 percent) (Figure 3–12). The white male rate remains the highest, but by 1970–75, it was only 13 percent higher than the nonwhite male rate.

The *all other* category includes the 13 major categories of cancer not included in any other major group. Although the diseases are a mixture, the results are prototypical. The white female rate was stable. The other three groups increased and the highest increase was for nonwhite males (Figure 3–13).

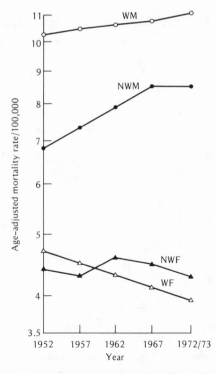

Figure 3–11. Trends in urinary system cancer mortality, 1950–75

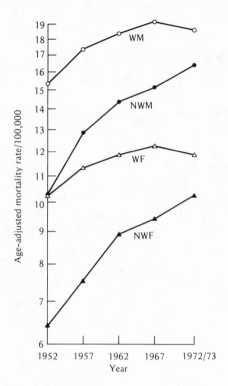

Figure 3–12. Trends in lymphatic system and hematopoietic cancer mortality, 1950–75

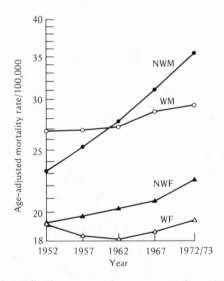

Figure 3–13. Trends in all other types of cancer mortality, 1950–75

Age-Specific Rates: All Types and Four Specific Types
Chosen to Illustrate Differences Between
Urban and Rural Areas

The section reviews age-specific rates for all types of cancer and lung, large intestine plus rectum, female breast, and bladder cancer mortality. The focus is on temporal change in the age-specific rates of the four populations. Accordingly, time series indices have been plotted as illustrations. The data are found in Appendix 1, Tables A1–9 through A1–13.

As will be shown in Chapter 4, the four types selected for presentation have historically exhibited a pronounced urban excess. By measuring the relative change of age-specific rates of these four and all types of cancer in urban and rural areas, it can be determined which age groups are contributing to convergence or divergence between urban and rural areas as well as which age groups are contributing to convergence and divergence among the four populations. The four types account for about 45 percent of cancer mortalities. The full 17 age groups and five time periods are presented for all types of cancer. Then, with one exception, age groups 35–85$^+$ and 1950–54 and 1970–75 are shown for the four specific types. The exception is bladder cancer, which has age-specific rates below 1/100,000 until the 45- to 49-year-old age group. The bladder cancer analysis accordingly begins with the 45–49 age group. The final note on data is that intestinal and rectal cancers were combined to lessen the death certificate accuracy problem noted in Chapter 2.

The age-adjusted rates of white males, white females, nonwhite males, and nonwhite females markedly changed during the study period. The age-specific rates of *all types* of cancer clearly identify the elderly age groups (65$^+$) as the populations most strongly influencing the changes in the age-adjusted rates (Figure 3–14; Table A1–9). The rates will be reviewed beginning with the youngest age groups. During 1950–54, the white male population had the highest child, teenage, and young adult cancer mortality rates (age groups 0–24), and the nonwhite female population usually exhibited the lowest rates. Cancer mortality rates for these age groups have gradually decreased. The most pronounced decrease was for the 0–4 age group, ranging from almost 50 percent for the white population to more than 30 percent for the nonwhite population (Figure 3–14).

The 5- to 14-year-old age groups are stable for nonwhite and show a 15 percent decrease for whites. All the 15- to 24-year-old age group rates decreased, especially the female rates. Overall, the 0- to 24-year-old pop-

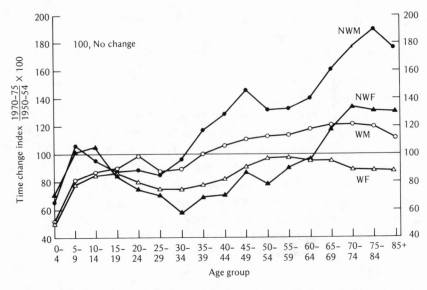

Figure 3–14. Change in all types cancer mortality, 1950–54 to 1970–75. Time change index of age-specific rates

ulation accounts for relatively few cases (2 percent of mortalities) and exhibits decreasing rates. Although the child, teenage, and young adult populations are not the major cause of the patterns noted in the age-adjusted rate section, the patterns exhibited by the younger age groups parallel the results for the entire population: white females manifested the largest decrease and nonwhite males the smallest decrease, which led to divergence between these populations.

However, the divergence of the male populations from the female populations and particularly the nonwhite male population from the other three populations is primarily due to relative rates of change in the elderly (65^+) and to a far lesser extent the 30–64 age groups. Nonwhite males manifested the highest rate of increase or the lowest rate of decrease in each of the 11 age groups from 30 to 85^+ years old. Nonwhite females exhibited the largest decreases for the 30- to 64-year-old population, and white females clearly manifested the largest decreases among the elderly (65^+).

With respect to cancer of the *lung,* every age-specific mortality rate of every subpopulation increased (Figure 3–15, Table A1–10). Nonwhite males began the 1950's with the highest rates for the 35–54 age population, about 10 to 15 percent higher than their white male counterparts.

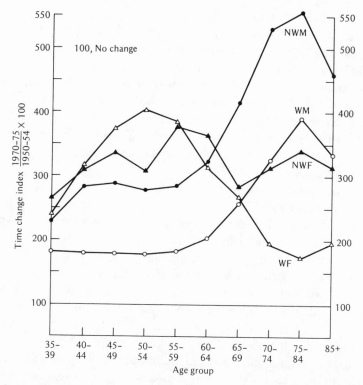

Figure 3–15. Change in lung cancer mortality, 1950–54 to 1970–75. Time change index of age-specific rates

While female rates were between one-fourth and one-seventh of the non-white male age-specific rates and were the lowest. White males 55 years old and older had the highest age-specific rates, particularly among the elderly. Nonwhite female rates were the lowest.

The rank order of the four populations was not greatly changed two decades later. However, actual differences between the populations were greater. Nonwhite male mortality rates increased far more rapidly than white male mortality rates in every age group (Figure 3–15). The 10 to 15 percent gaps between nonwhite and white males observed for those 35 to 54 years old during 1950–54 gradually increased from 75 to 90 percent by 1970–75. Among the 55- to 64-year-old population, nonwhite male rates tripled and white male rates doubled, leading to higher nonwhite than white male rates during 1970–75.

The largest increases for both white and nonwhite males were among those at least 65 years old. White male rates tripled and remained higher than nonwhite male rates. However, nonwhite male rates quadrupled and were almost as high as white male rates by the mid-1970's.

In contrast to the pronounced differences in rates of change between the male subpopulations in almost every age group, the female populations exhibited relatively similar rates of change for age groups 35–69 (Figure 3–15). There is however, a marked difference in the 70$^+$ age groups. Nonwhite female rates more than tripled compared to doubling among white females. It is this difference between the elderly age groups that accounts for the widening of the gap between the female age-adjusted cancer mortality rates.

Another noteworthy observation is the break that occurs at age group 60–64. Females exhibited the highest increases in the younger age groups, males in the older age groups. One cannot help but wonder whether this pronounced break is related to the observation that high female cigarette smoking rates lagged behind those of males by several decades. This observation will be considered later.

In conclusion, lung cancer is consistently a major contributor to the noteworthy divergence between nonwhite and white male cancer mortality rates. Lung cancer rates increased more rapidly among females than white males for the 35- to 69-year-old age groups. This convergence is contrary to the divergence between white males and females for all types of cancer. Lung cancer does, however, contribute to divergence between white males and the female populations for the 70$^+$ age groups.

Unlike most other major types of cancer, there is not much difference between the four populations in *large intestine* and *rectal cancer* mortality rates (Figure 3–16, Table A1–11). Nevertheless, some interesting changes occurred during the study period. Nonwhite females had the highest and either nonwhite or white males the lowest cancer mortality rates from these diseases during 1950–54 among those 35 to 59 years old. White males had the highest and nonwhites the lowest age-specific rates among the elderly age groups. During the subsequent two decades, white population rates decreased, while nonwhite rates increased. Thus, by the early 1970's, nonwhite rates had substantially increased relative to white rates (Figure 3–16). For example, nonwhite male rates for the 35- to 39-year-old age group started and ended at 3.9/100,000. Age-specific rates of the other three populations decreased about 35 percent. The result was that nonwhite males increased in rank from third (out of four) to first for this age group. By 1970–75, male or female nonwhites had the highest age-specific rates for every age group 35–59.

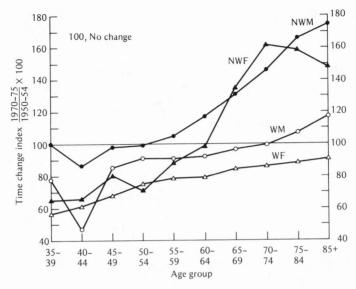

Figure 3–16. Change in large intestine and rectum cancer mortality, 1950–54 to 1970–75. Time change index of age-specific rates

White males exhibited the highest rates among the elderly throughout the study period. The gap between white males and the two nonwhite populations noticeably narrowed. During 1950–54, white females had the second or highest age-specific rate for every age group 60–85$^+$. Two decades later, due to decreases of 10 to 20 percent, white females had the lowest rates in the 60- to 74-year-old age groups.

Thus, nonwhite males had the lowest age-adjusted large intestine and rectum cancer mortality rates when the study period began and the second highest when it ended. Conversely, white females moved from the second highest to the lowest. This change is again primarily due to the elderly age groups.

White *female breast* age-adjusted cancer mortality rates were 22 percent higher than nonwhite female rates during 1950–54. By 1970–75, the gap had narrowed to 8 percent due to much higher increases among nonwhite females in the elderly age groups (65$^+$) (Figure 3–17, Table A1–12).

Cancer of the *bladder*, the last of the four specific types to be reviewed, has the lowest mortality rates. During the early 1950's, nonwhite males had the highest mortality rates for the 45- to 59-year-old age groups and

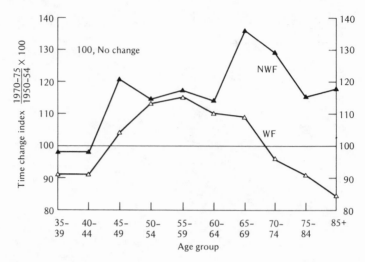

Figure 3–17. Change in female breast cancer mortality, 1950–54 to 1970–75. Time change index of age-specific rates

white males the highest rates for those at least 60 years old (Figure 3–18; Table A1–13). Two decades later, the rankings had not markedly changed. Nonwhite males exhibited the highest rates for the 45- to 64-year-old age groups and white males the highest among the elderly.

White females had the lowest age-specific rates for the 45- to 64-year-old age groups, by far, and nonwhite females for the elderly age groups. These rankings were similar to those two decades later.

The observation that the rank orders were quite stable during the study period understates some noteworthy changes, particularly the widening of the gap between male and female rates and the narrowing of the gap between nonwhite males and white males (Figure 3–18). In short, once again white females and nonwhite males were at the opposite ends of the spectrum with respect to changes in cancer mortality.

In summary, analysis of the age-specific rates for the all types and four selected disease types leads to the conclusion that the elderly age groups are responsible for the patterns discussed in the age-adjusted rate section. Rates are highest among the elderly, and these high rates reflect the largest increases or the smallest decreases. The most obvious change is the end of the plateaus that tended to characterize the nonwhite age-specific rate distributions beginning at ages 65–69 (Burbank, 1971). High rates of increase particularly characterize nonwhite males and are primarily responsible for the divergence of this population from the other three pop-

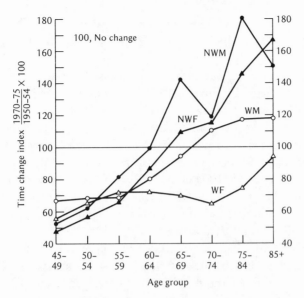

Figure 3–18. Change in bladder cancer mortality, 1950–54 to 1970–75. Time change index of age-specific rates

ulations. White females began the study period with the lowest age-adjusted rates and exhibited the smallest increases or the largest decreases in every age group, especially the elderly age groups. The result has been a widening of the gap between the white female population and the other three populations.

A COMPARISON OF CANCER MORTALITY RATES AND TRENDS IN THE UNITED STATES AND OTHER NATIONS

Comparisons of cancer incidence and mortality rates in different countries are fraught with problems regarding data such as differences in medical diagnosis and in recording the cause of death. Real differences relate to variations in diet, social customs, occupations, climate, medical practices, and genetic and many other factors that lead to differences in death rates. Despite wide variations in risk factors and record keeping, however, comparative cancer studies have been too interesting and revealing to neglect. It is important to provide an international perspective in order to avoid "not seeing the forest for the trees." For example, nonwhite and white male prostatic cancer mortality rates clearly diverge in the

United States. Which, if either, population is deviant? Lung cancer mortality rates have rapidly increased. Have they increased as rapidly in other nations?

International cancer studies have a long history. Among the first studies was one made by Hoffman (1915), who was engaged in a controversy as to whether cancer rates were increasing in Western nations. Hoffman reviewed data from the United Kingdom, Norway, Holland, Prussia, Switzerland, Austria, Australia, and New Zealand and from selected cities in Denmark and the United States. He concluded that there was an increase from about 45/100,000 in 1881 to 90/100,000 by 1911. He contended that "the evidence is so convincing that it may safely be maintained that no other statistical conclusion in medicine is so concisely and incontrovertibly established as this . . ." (p. 40).

The early comparative cancer studies are more interesting as anecdotes than as base-line data because of data limitations and the much smaller order of magnitude of the problem 50 to 100 years ago. For example, Hoffman offered the following report about cancer mortality trends in Frankfurt for the period 1906–13. Cancer mortality rates increased 10.8 percent for males and 8.6 percent for females from 1906–09 to 1910–13. Cancer of the digestive organs were responsible for almost 80 percent of the male deaths, and cancer of the reproductive and digestive organs for 80 percent of female deaths. With one exception, mortality was reported to have increased in every one of the major categories: cancer of the digestive system, skin, urinary, and reproductive systems, other carcinoma, sarcoma, and other malignant tumors. The one exception was the respiratory system, which Hoffman labels from the 1915 perspective as "relatively unimportant."

A Comparison of 1973 Cancer Mortality Rates in 52 Nations

The only international cancer mortality data available for comparative purposes was prepared by Segi et al. (1963, 1970, 1978) for the period 1950–73. From 1950 to 1967, cancer mortality of selected types in 24 countries was reviewed. The health statistics bureau of each country provided the data. Six reports were issued. Subsequently, Segi expanded the number of countries to 43 for the period 1970–72. Beginning in 1973, age-adjusted rates were prepared for 52 countries for the following 18 types of cancer using the World Health Organization A-classification:

1. Buccal cavity and pharynx
2. Esophagus

3. Stomach
4. Intestine, except rectum
5. Rectum, and rectosigmoid junction
6. Larynx
7. Trachea, bronchus, and lung
8. Bone
9. Skin
10. Breast
11. Cervix uteri
12. Other uterus
13. Prostate
14. Other and unspecified sites
15. Leukemia
16. Other lymphatic and hematopoietic
17. Benign and unspecified
18. All sites

Sixteen of the 18 types (excluding other and unspecified sites and benign and unspecified tumors) are of interest.

The standard population for the 52-country study, developed by Doll for the "world" population (Waterhouse et al., 1976), is reported below as percentages:

Age	% of Pop.	Age	% of Pop.
0–	2.4	40–44	6.0
1–4	9.6	45–49	6.0
5–9	10.0	50–54	5.0
10–14	9.0	55–59	4.0
15–19	9.0	60–64	4.0
20–24	8.0	65–69	3.0
25–29	8.0	70–74	2.0
30–34	6.0	75–84	1.5
35–39	6.0	85+	0.5
			Total 100.0

"African" and "European" standard populations have also been developed, but they are not used here.

According to Segi (1978), the standard population for 1973 is

slightly different from the ones used before for the statistics in 24 countries or in 43 countries. However, the age-adjusted death rates calculated this time seem to be practically comparable with the ones calculated and published up to the present.

Accordingly, it was decided to treat the 1950-73 data set as a time series, albeit one to be used with great caution. Specifically, after making sure that they were representative, 1952-53 and 1973 were chosen to represent the beginning and the end of the study period. When 1952-53 data were unavailable, the next data point, usually 1954, was used. Other Segi reports were used to verify that 1952-53 and 1973 were not unrepresentative years. Separate nonwhite and white U.S. rates were available for 1952-53, but not for 1973. Nonwhite and white rates for 1973 were calculated by the author using the Doll "world" population as the standard.

Before comparing the United States to the other 51 nations, the extent to which the 52 nations represent the world will be considered. They have a population of over one billion, but only about 27 percent of the world's population. The seven most populous of the 52 (United States, Japan, West Germany, Italy, France, England and Wales, and Mexico) each have over 50 million people, but are dwarfed in population by China and India. One-half of the 52 are located in Europe (Table 3-3); 15 are in Latin America, but Brazil and Argentina are not included. Africa is only represented by Mauritius.

The 1952-53 data are even less representative of the world. Sixteen of the 24 nations are in Europe. The remaining eight have been strongly influenced by European culture and European forms of urbanization and industrialization. Thus, the 52 nations are not a representative sample of the world's nations. Neither are they homogeneous in risk factors relevant to cancer. Changes in cancer mortality rates in many non-Western and non-urbanized industrialized nations may have little to do with exposure to carcinogenic agents, survival programs, and other cancer risk factors. Rather, cancer rates and trends in these non-Western nations may reflect nutritional improvements, advances in disease surveillance, water purification practices, insect control, and advances in non-chronic disease medicine (Preston, 1976). All the above practices prevent parasitic and infectious diseases. Accordingly, the comparisons that follow must be viewed as between the United States and two species of countries: (1) Western, urban-industrial and (2) non-Western and usually less industrialized nations.

The comparison leads to two conclusions. With some important exceptions, among the Western, urban nations, the United States usually has one of the lowest cancer mortality rates, but rates in the United States are almost always higher than rates for non-Western nations.

Beginning with *all sites,* the United States ranks twenty-second for males and twenty-third for females out of 52 nations (Table 3-4). Sev-

Table 3–3. Countries Included in the International Comparison

Country	1952–53	1973	Country	1952–53	1973
AFRICA			*OCEANIA*		
1. Mauritius		X	26. Australia	X	X
2. Union of South Africa	X		27. New Zealand	X	X
			EUROPE		
THE AMERICAS			28. Austria	X	X
			29. Belgium	X	X
3. Canada	X	X	30. Bulgaria		X
4. Chile	X	X	31. Czechoslavakia		X
5. Costa Rica		X	32. Denmark	X	X
6. Cuba		X	33. Finland	X	X
7. Dominican Republic		X	34. France	X	X
8. Ecuador		X	35. Fed. Rep. Germany	X	X
9. El Salvador		X	36. Greece		X
10. Honduras		X	37. Hungary		X
11. Mexico		X	38. Iceland		X
12. Nicaragua		X	39. Ireland	X	X
13. Panama		X	40. Italy	X	X
14. Paraguay		X	41. Luxembourg		X
15. Puerto Rico		X	42. Netherlands	X	X
16. Trinidad & Tobago		X	43. Norway	X	X
17. United States	X	X	44. Poland		X
18. Uruguay		X	45. Portugal	X	X
19. Venezuela		X	46. Romania		X
			47. Spain		X
ASIA			48. Sweden	X	X
			49. Switzerland	X	X
20. Hong Kong		X	50. England & Wales	X	X
21. Israel	X	X	51. N. Ireland	X	X
22. Japan	X	X	52. Scotland	X	X
23. Philippines		X	53. Yugoslavia		X
24. Singapore		X			
25. Thailand		X			

enteen of the 21 countries with higher male rates than the United States are European. The exceptions (Uruguay, Hong Kong, New Zealand, and Australia) are highly urbanized and have been strongly influenced by Europe. Eleven of the 30 countries with lower rates than the United States are located in Europe; but nearly all are less urbanized than the United States. The 15 lowest male cancer rates are found in Latin American, Asian, and African nations.

Table 3–4. Comparison of Age-Adjusted Cancer Mortality Rates in the United States and 51 Other Countries, 1973

Type	U.S. Age-Adjusted Rate/100,000	U.S. Rank	Five Highest Nations	Five Lowest Nations
1. *All sites*				
Male	161.38	22	Luxembourg	Nicaragua
			Czechoslovakia	El Salvador
			Scotland	Thailand
			Belgium	Honduras
			Netherlands	Dominican Republic
Female	108.96	23	Chile	Thailand
			Denmark	Nicaragua
			Hungary	Dominican Republic
			Scotland	Philippines
			Uruguay	El Salvador
2. *Lymphatic and other hematopoietic (excl. leukemia)*				
Male	9.06	2	Israel	Honduras
			United States	Thailand
			New Zealand	Nicaragua
			Denmark	Mauritius
			Sweden	El Salvador
Female	5.95	3	Isreal	Honduras
			New Zealand	Mauritius
			United States	Thailand
			Australia	Nicaragua
			Canada	El Salvador
3. *Large intestine*				
Male	14.24	7	New Zealand	Honduras
			Ireland	El Salvador
			Canada	Nicaragua
			Scotland	Dominican Republic
			Northern Ireland	Thailand
Female	12.51	9	New Zealand	Nicaragua
			Uruguay	Honduras
			Canada	El Salvador
			Ireland	Thailand
			Northern Ireland	Dominican Republic
4. *Lung*				
Male	50.38	10	Scotland	Nicaragua
			Eng. & Wales	Honduras
			Netherlands	El Salvador
			Luxembourg	Ecuador
			Belgium	Dominican Republic
Female	11.24	7	Hong Kong	Nicaragua
			Scotland	El Salvador
			Cuba	Honduras
			Eng. & Wales	Dominican Republic
			Singapore	Thailand
5. *Leukemia*				
Male	7.06	2	Greece	Thailand
			United States	Honduras
			Paraguay	Dominican Republic
			Northern Ireland	Nicaragua
			France	Trinidad & Tobago

Type	U.S. Age-Adjusted Rate/100,000	U.S. Rank	Five Highest Nations	Five Lowest Nations
5. *Leukemia (Continued)*				
Female	4.31	18	Denmark	Thailand
			Italy	Mauritius
			Canada	Nicaragua
			Netherlands	Dominican Republic
			Paraguay	Honduras
6. *Skin*				
Male	2.44	8	Australia	Nicaragua
			Norway	Honduras
			New Zealand	Mauritius
			Finland	Thailand
			Switzerland	El Salvador
Female	1.43	17	Iceland	Nicaragua
			New Zealand	Thailand
			Australia	Honduras
			Denmark	El Salvador
			Hungary	Panama
7. *Buccal cavity and pharynx*				
Male	4.71	11	Hong Kong	Honduras
			France	El Salvador
			Singapore	Iceland
			Puerto Rico	Nicaragua
			Luxembourg	Greece
Female	1.55	11	Hong Kong	El Salvador
			Singapore	Honduras
			Philippines	Nicaragua
			Iceland	Greece
			Venezuela	Mexico
8. *Female breast*				
	22.78	13	Scotland	Honduras
			Netherlands	Nicaragua
			Eng. & Wales	Thailand
			Ireland	El Salvador
			Denmark	Dominican Republic
9. *Prostate*				
	14.18	14	Switzerland	Thailand
			Sweden	Nicaragua
			New Zealand	Honduras
			Uruguay	El Salvador
			Australia	Singapore
10. *Esophagus*				
Male	4.04	25	Singapore	Nicaragua
			Puerto Rico	Honduras
			Uruguay	El Salvador
			France	Philippines
			Hong Kong	Dominican Republic
Female	1.14	24	Chile	Nicaragua
			Singapore	Honduras
			Ireland	El Salvador
			Puerto Rico	Thailand
			Uruguay	Dominican Republic

Table 3–4. Comparison of Age-Adjusted Cancer Mortality Rates in the United States and 51 Other Countries, 1973 (*Continued*)

Type	U.S. Age-Adjusted Rate/100,000	U.S. Rank	Five Highest Nations	Five Lowest Nations
11. *Larynx*				
Male	2.26	30	France	Nicaragua
			Luxembourg	Philippines
			Uruguay	El Salvador
			Spain	Thailand
			Italy	Honduras
Female	0.28	28	Venezuela	Luxembourg
			Iceland	Nicaragua
			Cuba	Honduras
			Ireland	Mauritius
			Puerto Rico	Sweden
12. *Cervix uteri*				
	4.56	31	Chile	Honduras
			Trinidad & Tobago	Spain
			Costa Rica	Thailand
			Venezuela	Nicaragua
			Mexico	Philippines
13. *Rectum*				
Male	4.39	30	Luxembourg	Honduras
			Czechoslovakia	Thailand
			Denmark	Nicaragua
			F.R. Germany	Paraguay
			Austria	El Salvador
Female	2.65	34	Denmark	Nicaragua
			Czechoslovakia	Honduras
			F.R. Germany	Thailand
			Hungary	Paraguay
			Austria	El Salvador
14. *Bone*				
Male	0.92	33	Luxembourg	Nicaragua
			Paraguay	Dominican Republic
			Romania	Honduras
			Iceland	Panama
			Portugal	Mauritius
Female	0.56	36	Paraguay	Nicaragua
			Romania	Mauritius
			Portugal	Honduras
			Greece	Thailand
			Italy	Norway
15. *Other uterus*				
	3.69	40	Paraguay	Nicaragua
			Mauritius	Hong Kong
			Venezuela	El Salvador
			Ecuador	Australia
			Trinidad & Tobago	Thailand
16. *Stomach*				
Male	7.47	47	Japan	Thailand
			Chile	Nicaragua
			Costa Rica	Dominican Republic
			Hungary	El Salvador
			Poland	Philippines

Type	U.S. Age-Adjusted Rate/100,000	U.S. Rank	Five Highest Nations	Five Lowest Nations
16. *Stomach (Continued)*				
Female	3.59	49	Japan	Thailand
			Chile	Dominican Republic
			Costa Rica	Nicaragua
			Portugal	United States
			Iceland	Philippines

The same European, urban/industrial and remainder of the world dichotomies hold for comparisons between female populations in the United States and other nations. Fifteen of the 26 European nations had higher female all types, cancer mortality rates than the United States in 1973; the United States exhibited higher female rates than all but six non-European nations. Five of the six exceptions were more urbanized than the United States in 1975 (Chile, Uruguay, New Zealand, Israel, and Venezuela). The sixth exception, Trinidad and Tobago, has extremely high rates of uterine cancer, which as will be seen in Chapter 4, is the only major type of cancer that usually tends to be higher in less urbanized, non-Western nations.

The United States clearly has among the highest rates for neoplasms of the *lymphatic* and *hematopoietic tissues,* excluding leukemia (Table 3-4, 2). White male rates are the second and female rates the third highest among the 52 countries. Only Israel had rates equal to or higher than those of the United States. The fact that Israel has higher rates than the United States is noteworthy because many persons of Eastern European, Jewish heritage have settled in the large cities of the United States (Haenszel, 1961; Greenwald et al., 1975). *Leukemia* mortality rates in the United States are also among the highest (Table 3-4). Among the populous, urban nations, Japan is conspicuous by its relatively low rates of leukemia as well as other lymphatic cancer. Urban areas in the United States with large concentrations of Japanese-born Americans should presumably have lower nonwhite rates of these types of cancer.

The United States has among the highest *large intestine* cancer mortality rates (Table 3-4). The countries with higher rates are Western nations located in Oceania (Australia, New Zealand) and in Northern Europe. Non-Western, subtropical and tropical nations exhibit the lowest rates. Among the urbanized nations, Japan and Poland have cancer mortality rates of the large intestine of only about one-third of the rate in the United States.

Japan and Poland, however, have among the highest *stomach cancer* mortality rates. The United States has among the lowest, about one-eighth of Japan's and one-fifth of Poland's. Dunham and Bailar (1968) add the Soviet Union, central Europe, and most of Latin America to the high rate and North America, Africa, and Southern and Southeast Asia to the low rate stomach cancer group. The difference between the United States and other nations, especially Poland and Japan, should be manifested in urbanized areas of the United States with many foreign-born Americans.

In contrast to large intestine and stomach cancer mortality rates in which the United States has either one of the highest or the lowest rates, *esophageal* and *rectal* cancer mortality rates in the United States fall about in the middle of the group of 52 countries, low compared to the European and high compared to the non-Western nations. The highest rates of cancer of the esophagus are primarily in urbanized Latin American and Asian nations (Singapore, Puerto Rico, Uruguay, Hong Kong). Dunham and Bailar (1968) also point to high rates in central Asia and parts of China. Among the populous European nations, only England and Wales and the adjacent nations (Scotland, Ireland, Northern Ireland) had markedly higher esophageal rates than those in the United States for both males and females in 1973. Nearly every European country had a higher rectal cancer mortality rate than the United States.

Given high levels of cigarette smoking and urbanization/industrialization, it is not surprising that the United States has among the highest *lung* cancer mortality rates. The nine nations with higher male rates than the United States are in Europe. Three of six nations with higher female rates than the United States are in Europe. Two others, Hong Kong and Singapore, are among the most urbanized nations in the world. The last, Cuba, is an enigma. It is a major tobacco producer, but production does not necessarily imply consumption. Moreover, male rates in Cuba are high, but not among the highest.

Although *trachea, bronchus,* and *lung* cancer mortality rates in the United States are relatively high, *larynx* rates are lower than in most of the 52 nations. The highest rates are found in France, North Africa, eastern Mediterranean countries, India, and some Latin American countries (Dunham and Bailar, 1968).

With respect to the reproductive organs, the United States has relatively high rates of *female breast* and *prostate* cancer and relatively low death rates from *uterus* cancer. The 12 nations with higher female breast cancer rates and the 13 nations with higher prostate cancer rates than the

United States are urbanized and strongly influenced by European culture. Among the urbanized nations, only Japan has low rates, ranking 45 of 52 in both types. Japan's female breast cancer rate in 1973 was only 21 percent of the rate in the United States; it's prostate rate was 15 percent of the rate in the United States. Non-western nations had the lowest rates, although Dunham and Bailar (1968) point to the USSR as a Western nation with relatively low female breast cancer mortality rates.

Uterus cancer mortality rates contrast strongly with most of the other types because urban, Western nations do not have the highest rates. Latin American nations have the highest rates, about four times the rates in the United States. European nations had among the lowest, but not the lowest. The lowest rates are usually found in Asia.

Buccal cavity and *pharynx, bone,* and *skin* cancers are relatively minor causes of mortality. The United States has among the highest buccal cavity and pharynx and skin cancer mortality rates and among the lowest bone cancer rates. Among the 52 nations, there is not a clear geographical pattern of buccal cavity and pharynx cancer. The highest and lowest rates are found in small Asian and European nations. Dunham and Bailar (1968) point to Southern and Southeast Asia and Latin America as having the highest rates and North America, Europe, and most of Africa as having the lowest rates.

The highest skin cancer mortality rates are found in nations with light-skinned populations, the lowest in Asian, African, and Latin American countries with dark-skinned populations. Finally, the highest rates of bone cancer mortality tend to be in southern European nations, the lowest in Latin American nations.

A Comparison of Cancer Mortality Trends in 23 Nations, 1952–53 and 1973

The United States was compared to 22 nations for the following types of cancer:

1. All types
2. Stomach
3. Intestine, except rectum
4. Trachea, bronchus, and lung
5. Female breast
6. All uterus
7. Prostate

The six specific types accounted for more than one-half of the cancer deaths in the United States during 1970–75. The Republic of South Africa was excluded from the comparison because 1973 data were not available. Other cancer sites were not included in the comparison because of data shortcomings. As a result of differences in recording of deaths, and the fact that the standard populations for 1952–53 and 1973 are different, precise trends cannot be calculated. However, general comparisons may be made using graphics (Figures 3–19 through 3–25).

Beginning with *all sites,* U.S. nonwhite males ranked sixth and U.S. white males ranked nineteenth of 24 populations in 1973 (Figure 3–19). These ranks reflect very different trends during the two decades. U.S. nonwhite males had among the highest increases. The increase from rank 14 in 1952–53 to six in 1973 was matched only by males in Belgium, Netherlands, France, and Italy. Like their nonwhite male counterparts, U.S. white males began the 1950's with an all sites rate in the middle of the set of 24 populations. Contrary to the U.S. nonwhite male population, white male rates increased only slightly during the study period. The result was a relative decrease from thirteenth to nineteenth among the 24 populations. Only Finland, Switzerland, Norway, and Israel had lower rates of increase.

One-half of the 24 female populations showed decreasing mortality rates from all types of cancer during 1952–53 to 1973 (Figure 3–19). Both U.S. female populations were among those that manifested decreasing rates. The U.S. nonwhite and white populations dropped from eighth to eleventh and from fourteenth to nineteenth, respectively. Iceland, Italy, and Portugal exhibited the largest increases.

The three major conclusions reached earlier in this chapter are generally supported by this initial international comparison of trends. Nonwhite male rates are not only diverging from the three U.S. populations, but in addition, nonwhite male rates in the United States have increased more rapidly than those in almost every other urbanized nation. Second, declining female rates are common. Third, the widening of the gap between male and female rates observed in the United States is an international phenomenon and not limited to the United States.

Earlier in this chapter, *stomach cancer* was found to be declining in all four U.S. populations. Stomach cancer mortality rates have also been decreasing in almost every other population (Figure 3–20). Portugal is the only nation among the 23 countries studied with increasing rates. Finland, Norway, and Switzerland manifested the largest absolute and

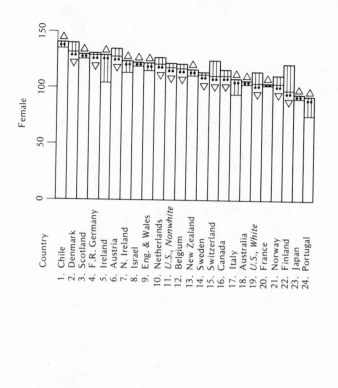

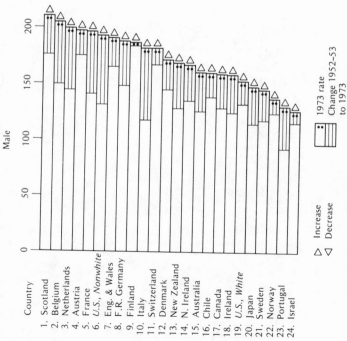

Figure 3–19. Comparison of all types cancer mortality rates in 23 countries, 1952–53 and 1973

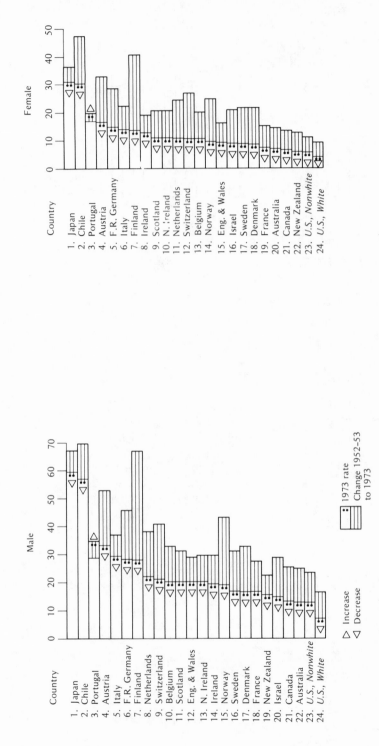

Figure 3–20. Comparison of stomach cancer mortality rates in 23 countries, 1952–53 and 1973

84

proportional decreases. The U.S. white populations had among the largest proportional decreases.

Also, earlier in the chapter, it was noted that nations that have high stomach cancer mortality rates tend to have low *intestine* cancer mortality rates. Japan, Chile, Portugal, and Finland have among the highest stomach and lowest intestine rates, whereas in the United States, Canada, and New Zealand the opposite is true.

The international trend provides an additional perspective about differences in large intestine cancer mortality rates in the United States. Earlier in the chapter, cancer of the large intestine was classified as one of six types exhibiting inconsistent trends in the U.S. populations because white female rates decreased while the rates of the other three populations increased. The international comparison suggests that the decrease in the U.S. white female population is part of an international trend toward convergence of female large intestine cancer mortality rates. Seven of the 24 female populations exhibited decreasing large intestine cancer rates during the study period (Figure 3-21). All seven had rates exceeding 10/100,000 during 1952-53. All nations that had rates below 10/100,000 during the early 1950's had higher rates during the 1970's, including U.S. nonwhite females.

Although there are signs of international convergence in the female large intestine data, most male rates increased. American white and nonwhite male rates were among the highest increases. They rose from the sixth and sixteenth ranks during 1952-53 to the fourth and ninth ranks two decades later, respectively. Only England and Wales and neighboring Scotland had decreases among the male populations.

Overall, what was seen earlier in the chapter as an inconsistent pattern in the United States seems to be part of an international pattern of change for cancer of the large intestine.

The fundamental question about *lung cancer* is whether other nations have exhibited the marked increases characteristic of the United States. The answer is a qualified "yes." Only Finnish females had a lower rate during the early 1970's than the early 1950's (Figure 3-22). The four U.S. populations showed higher increases than the populations of most nations. The U.S. nonwhite male rates increased from thirteenth to sixth in rank. Only males in Scotland, the Netherlands, and Belgium had increases equivalent to those for U.S. nonwhites. United States white males increased one rank, from tenth to ninth as the result of a proportional rate of increase that was higher than most nations. United States

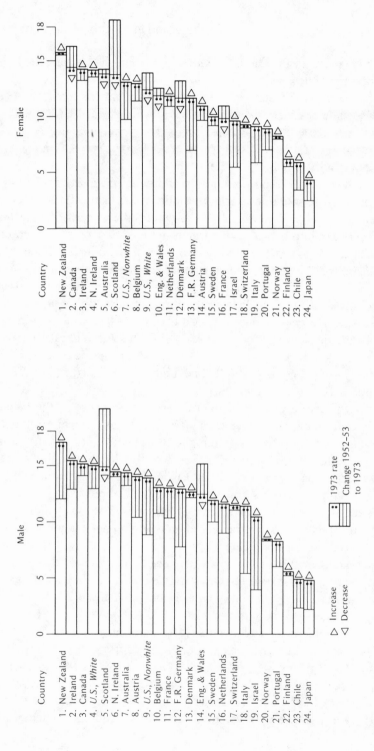

Figure 3–21. Comparison of intestine, excluding rectum, cancer mortality rates in 23 countries, 1952–53 and 1973

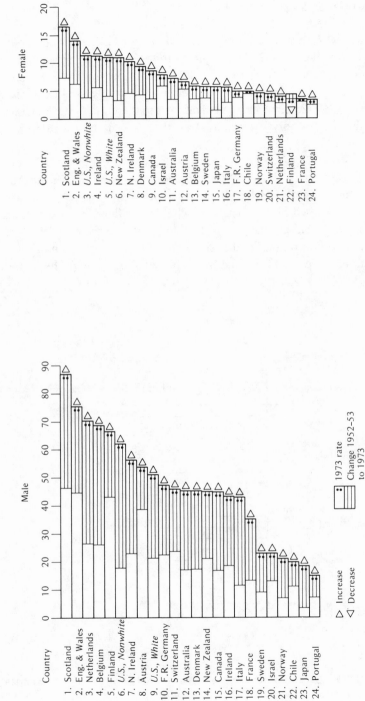

Figure 3–22. Comparison of trachea, bronchus, and lung cancer mortality rates in 23 countries, 1952–53 and 1973

female rates also increased more rapidly than rates in most other countries: white females increased from the tenth to the fifth highest and nonwhite females from the twelfth to the third highest. Overall, the dramatic increases that characterized lung cancer mortality rates in the United States from 1950–75 are clearly not confined to the United States; nearly every nation has rapidly increasing rates, though usually less than those in the United States.

Female breast cancer mortality rates have been relatively stable in most nations (Figure 3–23). Every population, with the exception of the Australian, which slightly declined, exhibited a slight increase. Rates of change in the United States were somewhat under those in most nations. In contrast to the slight increases of female breast cancer, rates of *uterus cancer* mortality decreased in almost every nation, in many by more than 30 percent (Figure 3–24). United States nonwhite rates were the highest

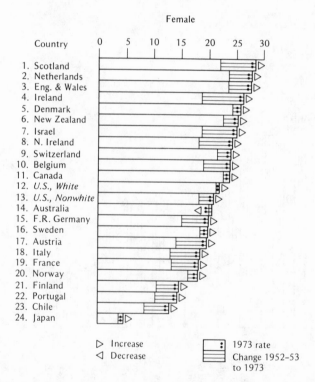

Figure 3–23. Comparison of female breast cancer mortality rates in 23 countries, 1952–53 and 1973

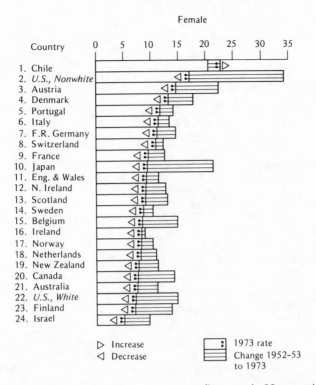

Figure 3–24. Comparison of uterine cancer mortality rates in 23 countries, 1952–53 and 1973

and U.S. white rates the seventh highest during 1952–53. Two decades later, despite pronounced decreases, nonwhite rates remained the second highest among the 24 populations, whereas white rates fell to twenty-second.

Earlier in the chapter, *prostate cancer* was placed in the inconsistent category because white male rates were relatively stable, whereas nonwhite male rates increased by more than 40 percent. The international data suggest that the two U.S. male populations fall at the opposite ends of the international continuum for prostate cancer. On the one hand, U.S. male nonwhites have the highest recorded rates, with among the highest increases in rate. On the other hand, U.S. white male rates are among the lowest and have been relatively stable. Clearly, the United States is an excellent place to study this disease.

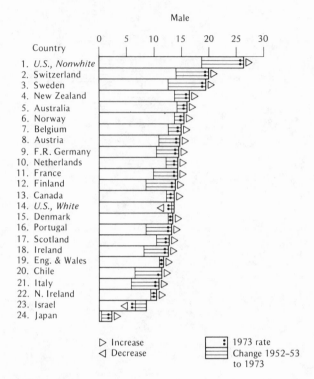

Figure 3–25. Comparsion of prostate cancer mortality rates in 23 countries, 1952–53 and 1973

SUMMARY

This chapter began with four questions, the answers to which will be briefly summarized:

1. Which types of cancer have become more important and which less important during 1950–75?

Seven types of cancer have become *far more important:* cancer of the lung and the pancreas, multiple myeloma, melanoma, cancer of the connective tissue, lymphoma, and cancer of the brain and central nervous system. *Five* types have become *slightly to moderately more important:* cancer of the female breast, ovary, larynx, nasopharynx, and kidney. *Eight* have become *far less important:* cancer of the stomach, rectum, liver, cervix uteri, corpus uteri, bone, other skin, and lip. *Eight* types have become *slightly to moderately less important:* Hodgkin's disease and cancer of the thyroid, nose, testis, male breast, eye, other endocrine, and salivary glands. *Six* types have been classified as *inconsistent* because the four populations vary in the extent to which the disease has changed during 1950–75: cancer of the esophagus and the prostate, leukemia, and cancer of the bladder, large intestine, and tongue and mouth.

2. Are cancer mortality trends in the United States different for white males, white females, nonwhite males, and nonwhite females?

Male rates are diverging from female rates. Male rates have been increasing since at least 1930, whereas female rates have been decreasing. Nonwhite males and white females are opposite. Although trends are generally similar for 28 of the 34 types, nonwhite males almost always have had the highest increases or lowest decreases; white females have usually exhibited the smallest increases or largest decreases.

3. In which age groups are the important cancer mortality trends in the United States occurring?

The elderly are responsible for the trends noted above. Their rates are not only the highest, but the elderly have generally manifested the largest increases or smallest decreases. High rates of increase among the elderly particularly characterize nonwhite males and are primarily responsible for the divergence between male nonwhites and the other three populations. White females exhibited the smallest increases or largest decreases in almost every age group for almost every type of cancer. The overall results has been a widening of the gap between white females and the other three populations.

4. How do cancer mortality rates and trends in the United States compare with other nations?

Age-adjusted cancer mortality rates for 16 types of cancer in 52 nations were compared for the year 1973. In addition, the trends between 1952–53 and 1973 were compared for 23 nations. The United States generally has lower rates than most Western, urban/industrial nations and higher rates than the non-Western and less industrialized nations that have higher mortality rates from infectious and parasitic diseases. Specifically, the United States has especially high rates of lymphatic and hematopoietic, large intestine, and lung cancer and low rates of stomach, uterus, and bone cancer. The trend analysis implies that the most important trends observed in the United States are common to Western, urban/industrial nations: the widening gap between males and females; the widening gap between nonwhite males and the other male populations; and decreasing female rates.

REFERENCES

Burbank, F. 1971. *Patterns in Cancer Mortality in the United States: 1950–1967,* NCI Monograph 33, Washington, D.C.: U.S. Government Printing Office.

Cutler, S., and Devesa, S. 1973. "Trends in cancer incidence and mortality in the U.S.A." In R. Doll, and I. Vodopija, eds. *Host Environment Interactions in the Etiology of Cancer in Man,* IARC Sci Pub. No. 7, Lyon, France: IARC, pp. 15–34.

Devesa, S., and Silverman, D. 1978. "Cancer incidence and mortality trends in the United States: 1935–74." *J. Nat. Can. Inst.* 60(3): 545–571.

Doll, R. and Peto, R. 1981. *The Causes of Cancer,* New York: Oxford University Press.

Dunham, L., and Bailar, J., III 1968. "World maps of cancer mortality rates and frequency ratios." *J. Nat. Can. Inst.* 41: 155–203.

Greenwald, P., Korns, R., Nasca, P., and Wolfgang, P. 1975. "Cancer in United States Jews." *Can. Res.* 35: 3507–3511.

Haenszel, W. 1961. "Cancer mortality among the foreign born in the United States." *J. Nat. Can. Inst.* 26: 37–132.

Hoffman, F. 1915. *The Mortality from Cancer Throughout the World,* Ch. 3. Prudential Press: Newark, N.J. pp. 28–47.

Lilienfeld, A., Levin, M., and Kessler, I. 1972. *Cancer in the United States,* Cambridge, Mass.: Harvard University Press, 1972.

Lilienfeld, A., and Lilienfeld, D. 1980. *Foundations of Epidemiology,* second edition, New York: Oxford University Press.

McKay, F., Hanson, M., and Miller, R. 1982. *Cancer Mortality in the United States: 1950–1977.* Washington, D.C.: Supt. of Docs.

Pollack, E., and Horm, J. 1980. "Trends in cancer incidence and mortality in the United States, 1969–76." *J. Nat. Can. Inst.* 64(5): 1091–1103.

Smith, S. 1980. "Government Says Cancer Rate is Increasing." *Science* 209: 998–1002.

Preston, S. 1976. *Mortality Patterns in National Populations.* New York: Academic Press, 1976.

Segi, M., and Kurihara, M. 1963. *Trends in Cancer Mortality for Selected Sites in 24 Countries 1950–1959,* Sendai, Japan: Tohoku University School of Medicine.

Segi, M., and Kurihara, M. 1970. *Cancer Mortality for Selected Sites in 24 Countries, 1964–1965.* Sendai, Japan: Tohoku University School of Medicine.

Segi, M. 1978. *Age-Adjusted Death Rates for Cancer for Selected Sites (A Classification) in 52 Countries in 1973.* Nagoya, Japan: Segi Institute.

United Nations 1980. Department of International Economic and Social Affairs, *Patterns of Urban and Rural Population Growth,* Population Studies No. 68, New York: United Nations.

Waterhouse. J., Muir, C., Correa, P., and Powell, J., eds. 1976. *Cancer Incidence in Five Continents, vol. III,* Lyon: IARC Scientific Publications, No. 15.

4 Changing Cancer Mortality Patterns in Urban and Rural America, 1950–75

The previous chapters discussed the thesis and methods and provided the baselines for comparing cancer mortality rates in urban and rural areas. This chapter answers three questions:

1. How do the actual urban/rural differences in cancer in the United States compare to differences that are suggested in the literature?
2. Is there an association between urbanization and cancer mortality rates in other nations?
3. How has the association between urbanization and cancer risk factors changed in the United States?

CANCER RATES AND DEGREE OF URBANIZATION IN THE UNITED STATES

The literature review in Chapter 1 suggests that some of the most important types of cancer should increase as urbanization increases. This section tests that hypothesis for the United States. Two comparisons are made. One is between two groups of counties that are at opposite ends of the urbanization spectrum: 32 counties that are completely or almost completely urban and 45 counties that were not urban in 1970. Then, due to the limited number of counties included in the first comparison, a far more detailed comparison is made between rural, moderately urban, and strongly urban counties.

After much deliberation, it was decided to limit the nonwhite analyses in this chapter to total cancer mortality rates. This decision was reluctantly reached, since there were sharp differences in cancer rates and geographical distribution of the four major nonwhite populations (American Indians, Japanese, Chinese, and black), as reported by Lilienfeld, Levin, and Kessler (1972) and Mason et al. (1976). Blacks, who comprise more than 90 percent of the nonwhite population, have high rates of cancer of the prostate. American Indians, who live mostly in rural areas, have much lower rates of cancer than the other nonwhite populations. Chinese-Americans, who nearly all reside in large urban centers, have very high death rates from cancer of the buccal cavity and pharynx, naso-pharynx, and biliary passages and liver. Japanese-Americans and Chinese-Americans have very high rates of cancer of the esophagus. Jap-anese-Americans, who live primarily in urban centers, have extremely high stomach cancer rates.

Depending upon the region of the United States, urban/rural com-parisons are slightly to strongly distorted by these sharp differences in cancer rates among Chinese, Japanese, and American Indians. In addi-tion, sharp differences between urban and rural-born blacks have been documented (Mancuso and Sterling, 1975; Mancuso, 1977). Similarly, Todd (1981) reported major differences between New York City blacks who were born in the United States and those who were born in Latin America. To supplement the nonwhite data given in Chapter 3, the author hereby provides a limited account of urban/rural differences for nonwhites in U.S. counties at the opposite ends of urbanization.

A Comparison between Counties at the Opposite Ends of the Urbanization Spectrum

According to the U.S. Bureau of Census (1973):

> The urban population comprises all persons living in (a) places of 2,500 inhabitants or more incorporated as cities, boroughs (except in Alaska), vil-lages, and towns (except in the New England States, New York, and Wis-consin), but excluding persons living in rural portions of extended cities (i.e., cities whose boundaries have been extended to include sizable portions of ter-ritory that is rural in character, such as city-county consolidations); (b) unin-corporated places of 2,500 inhabitants or more: and (c) other territory, incor-porated or unincorporated, included in urbanized areas (a central city, or cities and surrounding closely settled territory).

Using this imperfect definition of urbanization, the author found 32 counties that were at least 98 percent urban in 1970 (Table 4–1). Almost

Table 4-1. Counties at Least 98 Percent Urban, 1970

1. Alameda, Calif.	17. Hennepin, Minn.
2. Los Angeles, Calif.	18. Ramsey, Minn.
3. Orange, Calif.	19. St. Louis City, Mo.
4. San Francisco, Calif.	20. Carson City, Nev.
5. San Mateo, Calif.	21. Bergen, N.J.
6. Denver, Colo.	22. Essex, N.J.
7. District of Columbia	23. Hudson, N.J.
8. Broward, Fla.	24. Union, N.J.
9. Dade, Fla.	25. Los Alamos, N.M.
10. Muscogee, Ga.	26. New York City, N.Y.
11. Cook, Ill.	27. Nassau, N.Y.
12. Marion, Ind.	28. Cuyahoga, Oh.
13. Orleans, La.	29. Philadelphia, Pa.
14. Baltimore City, Md.	30. Dallas, Tex.
15. Suffolk, Mass.	31. Arlington, Va.
16. Wayne, Mich.	32. Milwaukee, Wisc.

900 counties had no urban population in 1970. A 5 percent random sample of the rural counties was taken. A comparison of total male and female white cancer mortality rates revealed no significant difference between this 45 county sample and the population of almost 900 counties with no urban population in 1970. Accordingly, the 45 counties were used to represent the population of rural counties (Table 4-2).

The 32 urbanized counties include some of the most populous American cities (Table 4-1). They had a population of about 40 million in 1970, compared to only 300,000 in the 45 rural counties (Table 4-2). The 45 counties had a nonwhite population of only about 45 thousand. To take into account the great disparity in size between the urban and rural counties, comparisons for whites were limited to the major disease aggregates reported in Chapter 3 and a few selected specific types. For nonwhites, comparisons were limited to the total cancer aggregate. In addition, comparisons were limited to 1950-69 because some of the rural counties may have become urbanized during the 1970's, a possibility that could not be checked because the 1980 census reports were not available at the time of this analysis.

Comparisons for 1950-69 as a single period revealed the expected: the completely urbanized counties have had much higher cancer mortality rates than the rural counties and the entire United States (Table 4-3), columns 3, 6; the rural counties have much lower rates than the entire United States and the urban counties (Table 4-3, columns 3, 7). The biggest difference was between nonwhite males in the urban and rural

Table 4–2. Sample Counties 0 Percent Urban, 1970

1. Lincoln, Ark.	24. Furnas, Neb.
2. Amador, Calif.	25. Hayes, Neb.
3. Archuleta, Colo.	26. Mora, N.M.
4. Routt, Colo.	27. Bertie, N.C.
5. Glascock, Ga.	28. Camden, N.C.
6. McIntosh, Ga.	29. Foster, N.D.
7. Taliaferro, Ga.	30. McLean, N.D.
8. Blaine, Idaho	31. Love, Okla.
9. Owen, Ind.	32. Custer, S.D.
10. Clark, Kan.	33. Somervell, Tex.
11. Kiowa, Kan.	34. Sutton, Tex.
12. Morris, Kan.	35. Garfield, Utah
13. Wichita, Kan.	36. Piute, Utah
14. Breckinridge, Kty.	37. Essex, Va.
15. Hancock, Kty.	38. Powhatan, Va.
16. Caldwell, La.	39. Russell, Va.
17. Nantucket, Mass.	40. Lincoln, Wash.
18. Crawford, Mich.	41. Pocahontas, W.Va.
19. Perry, Miss.	42. Florence, Wisc.
20. Christian, Mo.	43. Marquette, Wisc.
21. McCone, Mont.	44. Pepin, Wisc.
22. Meagher, Mont.	45. Niobrara, Wyo.
23. Dixon, Neb.	

counties (80 percent during the two decades). As noted above, the non-white male comparison is somewhat misleading because the nonwhites in many of the rural counties are not black, whereas the vast majority of those in the urban counties are black.

Differences between the rural and urban counties have decreased (Table 4–3, columns 4 and 5). By 1965–69, the difference between the total cancer mortality rates in the completely urban counties and the rural counties was about one-half of what it had been almost two decades earlier.

Comparisons of specific types also support the hypotheses drawn from the literature (Figure 4–1, Table A1–14). White male rates of respiratory, bladder, all urinary, large intestine, and all digestive system cancer were more than 50 percent higher in the completely urban counties than in the rural counties during 1950–69. Lymphatic and the all other type cancer aggregate were about 20 percent higher. Reproductive organ cancer mortality rates were about the same in the urban and rural counties.

The differences are less marked for white females than for white males, but are most pronounced for the expected organs. Female breast and respiratory cancer rates are more than 45 percent higher in the urban

Table 4–3. Comparison of All Cancer Types, Cancer Mortality Rates in Completely Urban, Rural, and All Counties in the United States, 1950–69

	(1)	(2)	(3)	(4)	(5)	(6)	(7)
			Rates/100,000				
			$\dfrac{\geq 98\% \text{ Urban}}{0\% \text{ Urban}} \times 100$			$\dfrac{(1)}{\text{U.S.A.}} \times 100$	$\dfrac{(2)}{\text{U.S.A.}} \times 100$
Population	≥98% Urban	0% Urban	1950–69	1950–54	1965–69		
White male	201.7	141.5	143	152	133	117	82
White female	144.9	116.8	124	125	116	111	90
Nonwhite male	225.9	125.6	180	196	154	123	69
Nonwhite female	155.2	114.3	136	155	124	110	82

97

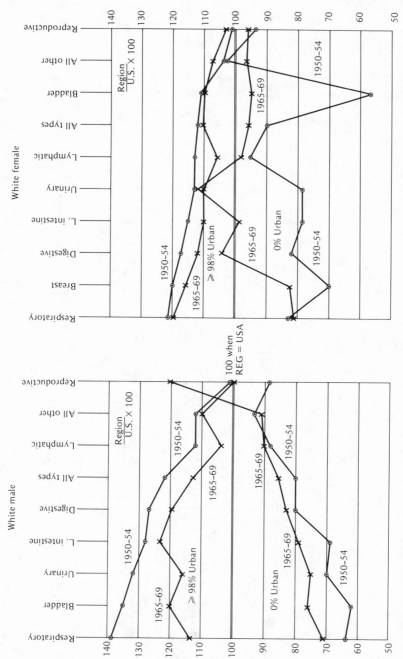

Figure 4–1. Comparison of age-adjusted cancer mortality rates in counties at opposite ends of the urbanization spectrum and the United States, 1950–69

than in the rural counties; bladder, digestive system, and lymphatic system rates are 20 to 25 percent higher. Only reproductive, total urinary, and the all other aggregate are less than 10 percent higher in the urban than in the rural counties.

Differences between the urban and rural counties have substantially narrowed for the major diseases (Figure 4–1; Table A1–14). In turn, the gap between these two extremes of American urbanization and the entire United States has narrowed (Figure 4–1).

As expected, the most pronounced differences between the most and the least urban counties in the United States are for the respiratory system, bladder, digestive system, especially the large intestine, and female breast cancer. Differences are far less marked for the lymphatic system, and are minimal for cancer of the reproductive organs and the all other types of cancer aggregate. Differences between the urban and rural counties have been shrinking. The cancer mortality profiles of both more closely resemble the profile of the total population of the United States during the late 1960's than during the early 1950's.

A Comparison of All Counties in the United States
Divided into Strongly Urban,
Moderately Urban, and Rural Aggregates

The above comparison of cancer mortality rates of counties at least 98 percent urban and 0 percent urban confirmed the literature-derived hypotheses that urban areas have substantially higher rates than rural areas of respiratory, digestive, especially the large intestine, female breast, and bladder cancer and added the observation that differences between urban and rural areas have been declining. Here the comparative analysis is broadened to include all the continental United States, with a finer breakdown of types of cancer, age-specific data, and 1970–75 data.

All counties in the continental United States were placed in one of three aggregates: strongly urban, moderately urban, and rural. Strongly urban counties were at least 75 percent urban in 1970. The almost 300 strongly urban counties contained 57 percent of the white population of the continental United States during the period 1950–75 (Table 4–4). Almost 20 percent of the white population lived in the more than 600 moderately urban counties. Between 50 and 74.9 percent of the population residing in these counties were classified as urban in 1970. Less than one-half of the population residing in the more than 2,000 rural counties

Table 4–4. Composition of Strongly Urban, Moderately Urban, and Rural County Groups

Aggregate	Counties: Continental U.S.	White Population 1950–75
Strongly urban	10%	57%
Moderately urban	21%	19%
Rural	69%	24%
Total	100%	100%

was classified as urban in 1970. About one-fourth of the white population resided in the rural counties in 1970.

A trichotomous classification of urbanization as a vehicle for comparing cancer rates is not new. Lilienfeld, Levin, and Kessler (1972) explored the relationship between urbanization and cancer mortality rates in the United States for 1959–61 with urban, suburban, and rural county aggregates. The urban counties were in metropolitan regions and contained a central city. The suburban counties were also in metropolitan regions, but did not contain a central city. The rural counties were not in metropolitan areas.

Although Lilienfeld, Levin, and Kessler's classification is different from the one used in this book, their results for 1959–61 are similar to those presented in this volume for 1960–64. Their urban/rural ratios for 1959–61 are close to the strongly urban/rural ratios for 1960–64. When their central city/suburban and suburban/rural ratios are compared to the strongly urban/moderately urban and moderately urban/rural ratios in this volume, their suburban aggregate has higher rates of cancer mortality than the moderately urban aggregate used in this volume.

A PROPORTIONAL MORTALITY PERSPECTIVE
Using types of cancer that account for at least 1 percent of cancer deaths, proportional cancer mortality data show some differences between the three aggregates during 1950–54. Beginning with *white males,* lung cancer was the most important cause of cancer mortality in the strongly and moderately urban counties, but only the third most important cause in the rural counties. Two decades later, lung cancer had almost doubled in importance and was the leading cause of all three aggregates (Tables A1–15, A1–16, and A1–17). Five other types exhibited obvious increases in in proportional mortality in all three aggregates: pancreatic cancer, non-Hodgkin's lymphoma, cancer of the kidney, melanoma, and multiple

myeloma. Six types of cancer exhibited decreases in order of importance: cancer of the stomach, rectum, liver, and esophagus, Hodgkin's disease, and non-melanoma skin cancer.

In addition to cancer of the lung, the three other most obvious differences between the urban and rural aggregates have also narrowed. Cancer of the prostate has become far less important in the rural areas, and by the early 1970's more closely resembled the proportional mortality rate of the urban aggregates than it did during the 1950's. The reverse is seen for the strongly urban areas for cancer of the esophagus and the larynx. These two types of cancer have become far less important in the strongly urban counties and slightly more important in the rural counties.

With respect to *white females,* lung cancer and to a lesser extent pancreatic cancer, non-Hodgkin's lymphoma, and multiple myeloma have become far more important causes of cancer mortality in all three aggregates. Lung and associated sites increased from between 3.2 and 3.8 percent of all female cancer-related deaths in 1950–54 to between 8.6 and 10.7 percent two decades later. Five other types decreased in proportional mortality in all three aggregates: cancer of the stomach, cervix uteri, corpus uteri, liver, and rectum (Tables A1–18, A1–19, and A1–20).

The proportional cancer mortality data did not distinguish as clearly between the strongly and moderately urban, on the one hand, and rural counties, on the other, for females as they did for males. The most urban counties had a larger percentage of breast, ovarian, and rectal cancer-related deaths than the least urban counties during 1950–54. The rural counties recorded a larger percentage of deaths due to cancer of the stomach, cervix uteri, corpus uteri, and liver than the most urban areas. The picture was even less clear two decades later. With one notable exception, nearly all the differences between the aggregates decreased. The exception is cancer of the trachea, bronchus, and lung. The proportional mortality rate increased more in the strongly and moderately urban areas than it did in the rural aggregate. This last observation is quite interesting and will be considered in more detail when the age-adjusted rates are presented.

In summary, when compared by proportional mortality, the urban and rural areas were similar with respect to the types of cancer that caused mortality during 1950–54. Cancer of the bladder (male), rectum and esophagus (male), larynx (male), breast (female), and ovary and lung were more important in the urban than the rural counties. Cancer of the stomach, prostate, cervix uteri, corpus uteri, and liver and leukemia (male), were more important in the rural than in the urban counties. By

1970–75, these urban/rural differences had substantially narrowed for almost every type of cancer.

A FINER PERSPECTIVE: COMPARING AGE-ADJUSTED
CANCER MORTALITY RATES

Proportional mortality is a useful indicator of the relative importance of a disease in a particular place, but it does not account for differences in the magnitude of rates between places. This section compares age-adjusted rates in strongly and moderately urban counties and rural counties. The major conclusion is that the white population of the nation is moving toward a homogeneous pattern of cancer mortality for almost every type of cancer. This conclusion is illustrated by Figures 4–2 through 4–6, and Tables A1–21 through A1–26.

Beginning with white males and all types of cancer, the white male rate in the strongly urban aggregate was 35 percent higher than the rate in the rural aggregate during 1950–54 (Table A1–21; Figure 4–2). It was 23 percent higher than the rate in the moderately urban counties and 12 percent higher than the national rate (Table A1–21; Figure 4–2). The differences declined during the next two decades. By 1970–75, the differences between the strongly urban aggregate and the other three—rural, moderately urban, and the entire United States—had fallen from 35, 23 and 12 percent to 12, 7, and 4 percent, respectively.

In comparison to white male rates, differences in white female rates between the most and the least urban aggregates were and continue to be less marked. During 1950–54, the total cancer mortality rate in the strongly urban aggregate was 16, 12, and 6 percent higher than the rates in the rural and moderately urban counties and the entire United States, respectively. Two decades later, the differences had decreased slightly to 12, 8, and 4 percent (Figure 4–3, Table A1–24).

A review of the 1950–54 data by type of cancer supports the literature-derived expectations. With respect to white male cancer, major differences between the strongly urban and rural aggregate were found for the expected respiratory, digestive and urinary system sites: bladder, kidney, larynx, lung, esophagus, rectum, tongue/mouth, and large intestine. The strongly urban/rural ratios for these eight sites averaged 1.91 and ranged from 1.39 to 2.43 (Table A1–21). Only multiple myeloma, with a strongly urban/rural ratio of 1.4, was a surprise because there is limited literature about urban/rural differences for this type of cancer (Anderson et al., 1970).

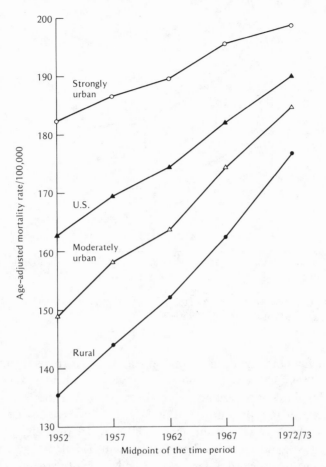

Figure 4–2. Trends in cancer mortality, white males: Strongly urban, moderately urban, and rural counties, 1950–75

   The lowest strongly urban/rural ratios were for the expected types: other skin cancer, cancer of the prostate, Hodgkin's disease, lymphoma, and bone cancer. Since exposure to sunlight is a key factor in non-melanoma skin cancer, rates should be and were higher in the rural than in the urban aggregate. Little is known about the factors contributing to cancer of the prostate. An urban/rural difference has not been reported (Hutchinson, 1976). Since prostate cancer is overwhelmingly a disease of the elderly, perhaps the disease has claimed a relatively large share of cancer deaths in rural areas because males in rural areas had lower rates

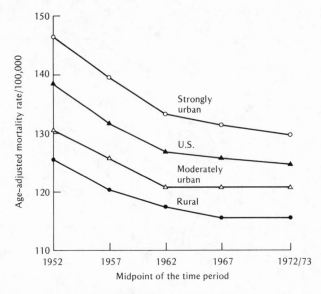

Figure 4–3. Trends in cancer mortality, white females: Strongly urban, moderately urban, and rural counties, 1950–75

of death for other types of chronic diseases. Or perhaps, prostate was listed as the cause because another cause was not obvious. Low urban/rural Hodgkin's disease and leukemia rates were expected insofar as there is a growing literature about risk factors for these diseases in rural areas (Adelstein, 1972; Nishiyama and Inove, 1970; Blair, 1979; Burmeister, 1980; and Donham, Berg, and Sawin, 1980).

Whereas urban/rural differences for white females are much less pronounced than they are for white males, most of the expected differences were found. Eleven strongly urban/rural ratios were at least 1.2 during 1950–54: rectal cancer, lymphoma, cancer of the ovary, breast, and bladder, multiple myeloma, cancer of the brain/central nervous system, Hodgkin's disease, and cancer of the lung, pancreas, and large intestine. The average strongly urban/rural ratio for these eleven was 1.31 in 1950–54. The lowest strongly urban/rural ratios were for cancer of the corpus uteri and liver, melanoma, and cancer of the cervix uteri (Table A1–24).

During the two and one-half-decade study period, the difference between the three aggregates dramatically decreased for white males and also decreased for white females (Figures 4–4 through 4–6). With respect to white males, the gap between the strongly urban and rural counties increased for only two of the 20 specific types: stomach and liver (Figures 4–4 and 4–5; Table A1–23). Both are declining causes of mortality. The narrowing of the gap is particulary apparent for the eight previously mentioned respiratory, digestive, and urinary types, which had an average ratio of 1.91 during 1950–54 (Figure 4–4 compared to Figure 4–5). The average, strongly urban/rural ratio for these same types decreased to 1.28 by the early 1970's. Finally, the unexpected urban/rural ratio for multiple myeloma during 1950–54 had changed by 1970–75, dropping from 1.4 to 1.0.

With respect to white females, the relatively small differences between the three aggregates further narrowed. The average, strongly urban/ rural ratio for the 11 types with the widest gap dropped from 1.31 to 1.14 (Figure 4–6; Table A1–26). Only one of these cancer types mani-

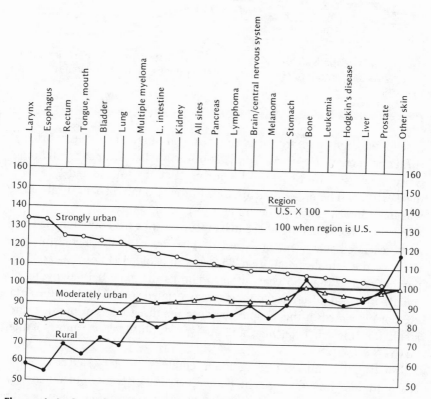

Figure 4–4. Comparison of white male age-adjusted cancer mortality rates: Strongly urban, moderately urban, rural counties, and the United States, 1950–54

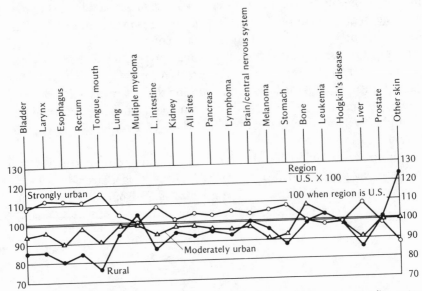

Figure 4–5. Comparison of white male age-adjusted cancer mortality rates: Strongly urban, moderately urban, rural counties, and the United States, 1970–75

fested a widening gap: trachea, bronchus, and lung. The strongly urban/ rural and moderately urban/rural ratios for trachea, bronchus, and lung cancer increased from 1.24 to 1.00 during 1950–54 to 1.38 and 1.13 during 1970–75. The only other increase in the strongly urban/rural ratios was for stomach cancer, a declining cause of mortality.

Frankly, the widening of the urban/rural ratios for white female lung and associated sites was not expected. The source of the increasing gap is the white female population 55 years and older (Figure 4–7). The population less than 55 years of age exhibits the expected narrowing of the urban/rural difference in lung and associated sites. Although the data are not presented in this chapter because the disease has a low rate, the result for trachea, bronchus, and lung is duplicated by cancer of the larynx.

Isolating the unusual finding about female respiratory cancer is much easier than is knowing why it has occurred. Differences in diagnostic practices, disease competition, and occupational exposures may be the answers. In the absence of any useful data, one cannot proceed further with these explanations. Air pollution and cigarette smoking offer the

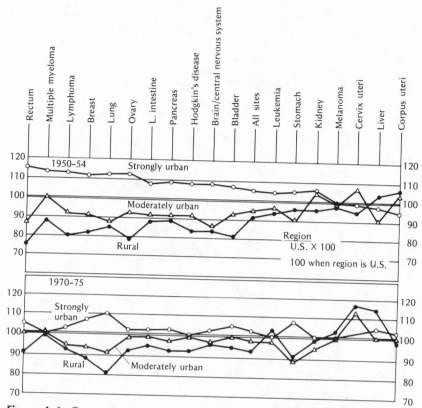

Figure 4–6. Comparison of white female age-adjusted cancer mortality rates: Strongly urban, moderately urban, rural counties, and the United States, 1950–54 and 1970–75

most interesting alternative explanations. If a latency period of at least 20 years is assumed for respiratory tract cancers, then contamination of urban atmospheres since the 1940's, which has been widely publicized, could be a contributing factor. Cigarette smoking is more likely to be accepted as an explanation, since females began to smoke later than males. If, as the literature and data show (see later section of this chapter), there was an obvious difference between smoking in urban and rural areas during the 1950's, which applied to women as well as to men, then, holding urban to rural migration constant, a widening of the gap between urban and rural areas in white female lung cancer mortality should have been expected. It follows, however, that the gradual erosion of the urban/

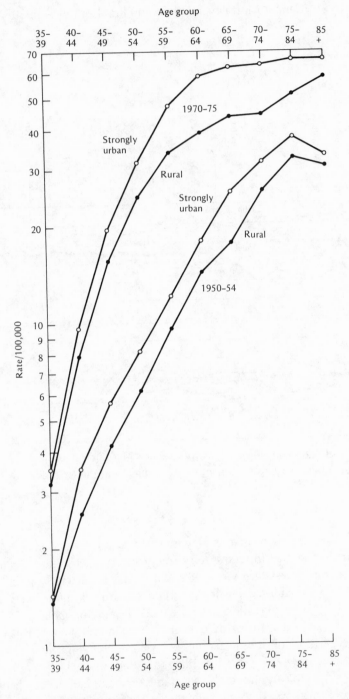

Figure 4–7. White female lung and associated cancer mortality rates, strongly urban and rural aggregates, 1950–54 and 1970–75

rural difference in cigarette use by the mid-1970's (see the last section of this chapter) should be manifested in a halt to and then perhaps a reversal of the widening urban/rural gap in white female lung cancer within the next two decades.

Whereas the difference between the most and least urbanized counties decreased, have the strongly urban, moderately urban, and rural counties manifested similar trends for each type of cancer? Or are there obvious differences between the three aggregates for many or a few types of cancer? Parallelism and deviation were tested by applying rank correlation methods to the time change indices for the three aggregates and the United States (Table 4-5; see Appendix 2 for a discussion of the application of the rank correlation statistic to these data).

There is a strong parallelism between the strongly urban, moderately urban, and rural aggregates and the entire United States. The rank correlation between each aggregate (e.g., white male strongly urban with white male moderately urban) range from .91 to .99 and are all significant at the .01 level (Tables A1-27 through A1-29). Clearly the diseases that have risen sharply or declined in the United States have also risen sharply or declined in each of the aggregates.

Although there is a strong tendency toward parallel change, seven types have noteworthy differences, especially between the strongly urban and rural aggregates: lung cancer; multiple myeloma; and cancer of the larynx (only male), esophagus (only male), bladder (only male), tongue/mouth (only male), and ovary. With the exception of female lung cancer, the rural county aggregate exhibited the largest increase, whereas the strongly urban aggregate manifested the larger decrease or a much smaller increase (Table 4-5). White male mortality rates of cancer of the lung and of multiple myeloma almost doubled in the strongly urban counties; but the rates of these two types nearly tripled in the rural counties during the study period. The more than 2,000 rural counties exhibited slight to moderate increases in white male cancer mortality rates of the larynx, esophagus, bladder, and tongue/mouth. Rates for these four types in the strongly urban counties declined.

The striking parallelism between the three aggregates and between male and female rates in rates of change, and the tendency toward a homogeneous geography of cancer because the rural rates almost always increase more than or decrease less than the most urban counties are the main observations seen in the age-adjusted rates. White female lung cancer stands out as an exception to the trend toward an equalized national geography of cancer.

Table 4–5. Change in Age-Adjusted Cancer Mortality Rates, Strongly Urban, Moderately Urban, Rural Counties and the United States, 1950–54 to 1970–75

| Site | $\dfrac{1970-1975}{1950-1954} \times 100$ | | | |
	SU[a]	MU[b]	RU[c]	U.S.A.
White Male				
1. Lung	204	280	324	239
2. Stomach	43	42	41	43
3. Prostate	95	98	99	97
4. Large intestine	109	121	127	115
5. Rectum	59	77	81	66
6. Pancreas	116	130	143	126
7. Leukemia	102	113	118	109
8. Bladder	89	107	118	100
9. Liver	52	47	45	49
10. Tongue, Mouth	84	103	107	91
11. Esophagus	82	109	143	98
12. Lymphoma	140	153	161	146
13. Brain/central nervous system	121	133	140	126
14. Kidney	115	139	150	129
15. Larynx	91	125	157	108
16. Hodgkin's disease	75	82	86	78
17. Other skin	50	47	50	47
18. Bone	56	65	56	59
19. Multiple myeloma	179	218	250	208
20. Melanoma	177	182	210	183
21. Total	108	124	130	116
White Female				
1. Breast	99	106	108	102
2. Large intestine	85	96	93	89
3. Stomach	39	38	37	39
4. Cervix uteri	44	49	57	47
5. Corpus uteri	53	47	46	49
6. Ovary	99	113	120	106
7. Liver	41	38	38	38
8. Rectum	51	65	67	56
9. Leukemia	90	96	104	95
10. Pancreas	112	122	121	117
11. Lung	275	279	248	274
12. Bladder	67	73	78	68
13. Lymphoma	138	154	171	150
14. Brain/central nervous system	125	143	150	131
15. Kidney	100	95	111	105
16. Hodgkin's disease	79	92	91	85
17. Melanoma	154	156	153	154
18. Multiple myeloma	189	213	243	196
19. Total	89	92	92	90

[a]Strongly urban counties (≥75% urban in 1970).
[b]Moderately urban counties (50 to 74.9% urban in 1970).
[c]Rural counties (<50% urban in 1970).

AGE-SPECIFIC RATE PERSPECTIVE

Comparison of age-specific rates in strongly urban, moderately urban, and rural counties and the entire United States provides additional insights about the trend toward homogeneity of cancer mortality rates in the United States. Beginning with *white males,* the strongly urban counties had higher rates than the entire United States for every one of the 30+-year-old age groups during the early 1950's (Table A1–30, Figure 4–8). The strongly urban/U.S.A. ratios have two peaks: one at age group

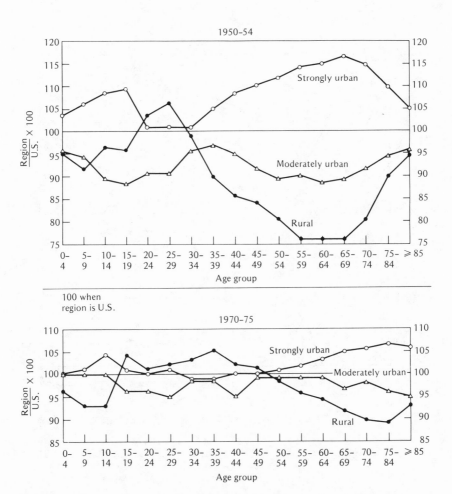

Figure 4–8. Comparison of white male age-specific cancer mortality rates, all sites, 1950–54 to 1970–75: Strongly urban, moderately urban, rural counties, and the United States

15–19 and a second, higher peak at age group 65–69. The rural/U.S.A. ratios are virtually a mirror image of the strongly urban/U.S.A. ratios. The rural/U.S.A. ratios are above 1.0 for age group 20–29. These two are the only age groups for which the strongly urban/U.S.A. ratios were less than 1.0 during 1950–54. The rural/U.S.A. ratios drop substantially below 1.0 at age group 35–39 and continue to drop to .76 at age group 55–59. In comparison to the strongly urban and rural aggregates, the moderately urban/U.S.A. ratios are closer to the U.S. totals ranging from .88 to .97 (Table A1–31).

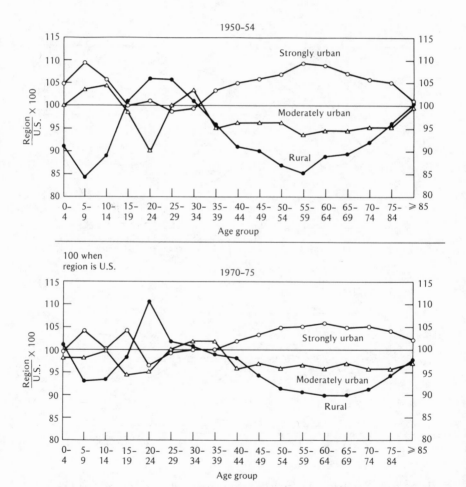

Figure 4–9. Comparison of white female age-specific cancer mortality rates, all sites, 1950–54 to 1970–75: Strongly urban, moderately urban, rural counties, and the United States

The 1970–75 graph of ratios for white males stands in strong contrast to the 1950–54 one (Figure 4–8). All three curves are much closer to the national totals. The initial peak (age group 5–19) for the strongly urban counties has almost disappeared and the stronger second peak is not as high and does not appear until age group 55–59. The moderately urban ratios range between .95 and 1.00. The rural ratios for the adult population all climbed much closer to the national totals.

The most important observation seen in the age-specific rate data is that by 1970–75 a noticeable difference between cancer mortality rates in the most urban and least urban counties did not exist until age group 55–59. Since two decades earlier an obvious difference had existed by age group 35–39, it follows that if the trend toward homogeneity continues at the rate implied in Figure 4–8, there will be no further difference between urban and rural aggregate rates by the years 1990–2000.

The white female data offer a less clear trend (Figure 4–9). Urban/rural differences were apparent by age group 35–39 during the early 1950's. Two decades later, they are not apparent until age group 45–49, and in general, the urban/rural gradients for white females have not decreased as much as they have for white males. If, however, cancers of the respiratory system are removed from the analysis, there is a much sharper trend toward homogeneity.

THE ASSOCIATION OF CANCER MORTALITY RATES AND URBANIZATION IN COUNTRIES OTHER THAN THE UNITED STATES

The literature review and the initial analysis of urban and rural United States data clearly point to marked, but at least in the United States, declining urban/rural differences in cancer of the respiratory, digestive, and urinary systems and in female breast cancer. Before smaller slices of the American data are examined, it is appropriate to review non-U.S. data for urban/rural differences. Two reviews are presented. One is of the association between urbanization and the 1973 cancer mortality rates for 52 nations. The second is of the urban/rural differences in Poland, Norway, Canada, and England and Wales.

The Association of Cancer Mortality in 1973 and the Degree of Urbanization in 52 Nations

If cancer rates increase with urbanization, then the nations with the highest cancer mortality rates in 1973 should be the most urbanized, the nations with the lowest rates, the least. A recent United Nations publication was used as an indicator of urbanization (1980), and the 52 nations were ranked according to degree of urbanization and then divided into three urbanization classes: high (ranks 1–17 in urbanization), middle (18–35), and low (36–52). The 52 countries were also divided into high, middle, and low cancer mortality rate groups for 16 types of cancer (data available for 1973) (Segi, 1978; see Chapter 3 for a discussion of the data). One noteworthy absence is urinary tract data.

Chi-square tests were performed on the data organized into 3 by 3 tables and evaluated by significance tests and the contingency coefficient. The closer the coefficient is to 1 or −1, the stronger the relationship; the closer to 0, the weaker.

Fifteen of the 28 tests produced results insignificant at the .05 level:

1. Male skin
2. Female skin
3. Male leukemia
4. Female leukemia
5. Male stomach
6. Female stomach
7. Male larynx
8. Female larynx
9. Male bone
10. Female bone
11. Male buccal cavity and pharynx
12. Female buccal cavity and pharynx
13. Female rectum
14. Male prostate
15. Female cervix

Using the literature and the U.S. data as sources of expectations, 12 of the 15 insignificant results were expected and three were unexpected (female rectum and male and female stomach cancer). The insignificant female rectal relationship was in the expected direction and was significant at the .25 level. The stomach cancer results imply no relationship between urbanization and stomach cancer mortality rates. Some of the most urbanized nations (Australia, New Zealand, Canada, the United States) had low stomach cancer mortality rates and among the highest large intestinal cancer rates. The converse was also true. Some of the least urbanized nations (Bulgaria, Costa Rica, Romania, Ecuador) have high stomach and very low large intestinal cancer mortality rates. Stomach cancer mortality, unlike large intestinal cancer mortality and some of the other digestive organ cancers, does not seem to be associated with degree of urbanization measured at the national level. The case studies below suggest that there is also little consistency in the association within nations. Thirteen of the 28 statistical tests produced significant associations at the .05 level or better (Table 4–6). Ten of the thirteen had been expected. The two strongest associations are between intestinal cancer and urbanization. For both males and females, 11 of the 17 most urbanized nations (out of 52 in all) also have the highest intestinal cancer mortality rates. No nations among the least urbanized have high intestinal cancer mortality rates.

The third strongest association is with female breast cancer mortality rates. None of the 17 highest rates for this disease were found among the least urbanized nations, whereas twelve of the 17 highest rates of female breast cancer mortality are found in the most urban nations. Two of the most urbanized nations, Hong Kong and Singapore, had among the lowest female breast cancer mortality rates. The added observation that the three other Asian nations (Japan, the Philippines, and Thailand) also had low rates suggests an interesting exception to the general tendency for a strong association between urbanization and female breast cancer mortality rates.

Table 4–6. Association of Urbanization and Cancer Mortality Rates for 1973 for 13 Types

Type	Sex	Contingency Coefficient[a]	Significance
1. Intestine	Female	.69	.001
2. Intestine	Male	.65	.001
3. Breast	Female	.64	.001
4. Other lymphatic	Male	.61	.01
5. Lung	Female	.61	.01
6. All types	Female	.57	.01
7. All types	Male	.56	.01
8. Esophagus	Male	.55	.02
9. Lung	Male	.53	.02
10. Rectum	Male	.53	.02
11. Other uterus	Female	−.52	.02
12. Other lymphatic	Female	.50	.05
13. Esophagus	Female	.49	.05

[a]Coefficient divided by maximum value of a 3 by 3 table so that the range is 1 to −1.

The fourth and twelfth strongest associations are with other lymphatic cancer mortality. These might be considered unexpected because there is not a large literature suggesting an urban/rural difference. The U.S. data disclosed an urban/rural difference, but one not nearly as strong as those for the respiratory and digestive system and female breast and bladder cancer.

There are five exceptions to the strong association of other lymphatic cancer mortality rates in 1973 and urbanization. Hong Kong, Singapore, and Spain are among the most urban nations, but have some of the lowest other lymphatic cancer mortality rates. Finland and Norway are among the least urban nations, but have some of the highest rates of these diseases. The observations that Japan, Thailand, and the Philippines also have among the lowest rates of these diseases and that Denmark and Sweden also have among the highest rates of other lymphatic cancer suggests that two different regional factors may be operating to modify the overall association of this disease with urbanization.

Given the amount of writing about the urban factor in lung cancer, perhaps it is surprising that the associations between lung cancer and urbanization at the national scale are only the fifth and ninth strongest. Hungary is the lone noteworthy exception among the 52 nations. There lung cancer mortality rates were higher than would have been expected from level of urbanization.

Esophageal and male rectal cancer mortality rates are also higher in the more urban nations. There were no obvious regional patterns among the exceptions. The overall relationship between urbanization and cancer mortality is summarized by the significant associations between all types of cancer and urbanization (Table 4–6).

The 28 tests found one clear exception to the positive association between urbanization and cancer mortality among the 52 nations. The greater the urbanization, the lower the rate of other uterine cancer mortality. Eleven of the 17 lowest rates were in the most urbanized nations: Hong Kong, Australia, New Zealand, Scotland, Israel, Singapore, England and Wales, Canada, the United States, Netherlands, and Sweden. Some of the least urban nations have among the highest rates: Trinidad and Tobago, Honduras, Italy, Mauritius, Ecuador, Romania, Paraguay, and Hungary.

Thus, these statistical analyses further strengthen the case for an association between the degree of urbanization and lung, some types of digestive system, female breast, and total cancer mortality rates. On the other hand, female other uterine cancer mortality rates tend to be higher in the least urbanized areas.

Urban/Rural Differences within Four Nations:
Poland, Norway, Canada, and England and Wales

This section considers the extent to which there is an urban/rural difference for different types of cancer within nations. Four countries are analyzed (Poland, Norway, Canada, and England and Wales) and compared
to the United States. Two of the four, England and Wales and Canada,
are among the most urban of the 52 nations and also have relatively high
cancer mortality rates. Poland and Norway rank lower in urbanization.
But only Norway has among the lowest cancer mortality rates as indicated in the following table.

Nation	Rank in Urbanization	Rank in Cancer Mortality, 1973	
		Male	Female
England and Wales	2	9	11
Canada	15	24	21
Poland	27	19	25
Norway	45	30	28
United States	17	22	23

POLAND

The most useful information came from Poland. Located in central
Europe, this nation of 33 million has been characterized by its relatively
high stomach and relatively low large intestine and female breast cancer
mortality rates (see above and Chapter 3; Segi, 1978). Straszewski's
excellent monograph (1976) explores these and other diseases for the
period 1959–69. Noting that Poland has been transformed from an agri-
cultural to an industrial country, Straszewski presents the case that
Poland is moving toward an urban Western cancer pattern. Stomach cancer rates are falling, whereas intestinal and female breast rates have
become more like those of other Western nations.

Of particular interest to this study are the urban/rural cancer mortality ratios for seven categories:

1. All sites
2. Stomach
3. Intestine and rectum
4. Lung
5. Female breast
6. Uterus
7. Prostate

Straszewski's plots of age-specific rates (five-year age groups) and age-adjusted rates clearly indicate changes in urban and rural areas between 1959–61 and 1967–69. During 1959–61, residents of urban areas in Poland had much higher cancer mortality rates than residents of rural areas, for every major type of cancer. Nearly all the difference was among those at least 55 years old. Beginning with all sites and 1959–61, Straszewski reports urban male and urban female rates 60 to 65 percent higher than rural male and rural female rates. Almost every rate was more than twice as high in urban areas. Among the major types, only stomach cancer had a ratio under 2.0. The relatively low urban/rural ratio for stomach cancer (1.21) is suggested as being primarily due to the association of the disease with a diet characteristic of rural poverty.

Less than one decade later, all the urban/rural ratios were lower. The all-sites rate declined from about 1.65 to 1.25 for both males and females, and stomach cancer rates were higher in rural than in urban areas. With respect to females, the urban/rural ratios for uterine, intestine, rectal, and lung cancer fell below 1.5. Only the female breast cancer urban/rural ratio continued above 1.5, but it had decreased to about one-half of what it had been during 1959–61.

Male urban/rural ratios had also decreased by 1967–69, but generally the decreases were more modest than the female ratios. Male urban/rural ratios of cancer of the intestine and rectum and lung fell to about 1.75, and the ratio for prostate fell to about 1.5.

Straszewski's analysis is supported by incidence data for 1968–72 (Waterhouse et al., 1976). A comparison was made between the City of Warsaw, the capital of Poland, with over one million residents, and rural Warsaw, with over one-half million residents. The following Warsaw City/rural Warsaw ratios are above 2.0: male and female intestine and rectum; male and female trachea, bronchus, and lung; female breast; female bladder; male and female urinary system. By comparison, male and female stomach cancer incidence rates in urban and rural areas were virtually identical.

As a summary indicator, a comparison was made of Warsaw City/rural Warsaw incidence ratios for two aggregates of cancer: (1) intestine and rectum; trachea, bronchus, and lung; female breast; bladder and other urinary, and (2) all other sites. City/rural ratios for the first group were 2.0 for males and 2.3 for females. The ratios for the remaining types of cancer were 1.3 for males and 1.5 for females.

Straszewski's study, combined with the incidence data, points to the expected marked difference between rural and urban areas for some specific types of cancer. Like the United States, a strong tendency toward more uniformity between urban and rural areas is noticeable. Straszewski points to improved data as well as urbanization and Westernization as the major causes of changing patterns of cancer mortality in Poland.

NORWAY

Located in northern Europe, Norway is a nation with lower cancer mortality rates than most Western nations. The only exceptions are prostate and skin cancer and leukemia and other lymphatic cancers, for which Norway ranked between second and sixteenth of 52 nations in 1973.

Urban/rural incidence data were available for 1968–72 (Waterhouse et al., 1976). As expected, urban rates were higher than rural rates, although they were not as different as they were for Poland. Excluding other skin cancer because of its extremely high incidence relative to mortality, urban/rural ratios were calculated for all sites and also for many specific sites. The all sites urban/rural ratio during 1968–72 was 1.3 for males and 1.2 for females. The aggregate of intestine and rectum; trachea, bronchus, and lung; female breast; bladder and other urinary, as expected, yielded higher ratios than for the all other sites aggregate. Specifically, the ratios for the first aggregate were 1.6 for males and 1.3 for females, compared to 1.2 for males and 1.1 for females for all other sites. The highest urban/rural ratios were for trachea, bronchus, and lung (about 2.0) and bladder (about 1.7). Although the Norwegian data are limited, they support the previous findings.

CANADA

Lying north of the continental United States, Canada has relatively high intestinal cancer and leukemia and other lymphatic cancer mortality rates and relatively low stomach and uterine cancer rates. The Bureau of Epidemiology (1980) in one of its two atlases of mortality for the period 1966–76, published in 1980, focuses on 14 specific types of cancer, all other, and all sites of cancer:

1. All sites
2. Tongue, mouth, and pharynx
3. Stomach
4. Large intestine, except rectum
5. Rectum
6. Large intestine and rectum

7. Pancreas
8. Lung
9. Female breast
10. Uterus
11. Ovary
12. Prostate
13. Bladder
14. Lymphatic tissue
15. Leukemia
16. Other cancers

The atlas provides data for 200 Canadian census divisions and is illustrated with 28 color maps. Although the data were not perfectly applicable for urban/rural comparisons, the association between urbanization and cancer mortality rates from the maps and the rates could be roughly inferred.

Red, orange, and yellow represent rates in the upper two deciles of census units. If there is an association between cancer mortality rates and urbanization, red, orange, and yellow for Canadian urban areas should be predominate. Unfortunately, because the use of the colors red, orange, and yellow also reflect statistical significance measures, many of the maps are almost all light green, the color representing "not significantly different from national rates." The majority of the maps did not suggest an urban excess. Largely because of the conservative manner of developing the maps, some of the maps showed almost no variation in cancer mortality rates across Canada. These included cancer of the tongue, mouth, and pharynx; female rectum; ovary; female bladder; and female leukemia. Higher rates in rural areas than in urban areas were seen for male leukemia and stomach and pancreatic cancer for males and females. Some of the highest rates of male rectum, female bladder, and male and female lymphatic cancer were found in cities (usually Montreal), but otherwise the maps did not suggest a consistent association with urbanization.

The maps suggest an urban/rural difference for the following types of cancer: male and female all sites, lung, and large intestine, female breast, and male bladder.

The highest total male cancer mortality rates are found in the Province of Quebec from Sept-Iles along the St. Lawrence Seaway to Hull in Ontario Province. Urban Montreal, Quebec, Winnipeg, Toronto, Hull, Chicoutimi, Chambly, and St. Maurice have among the highest rates. The prairie provinces exhibit the lowest rates. The largest prairie and

Western cities, Calgary, Edmonton, and Vancouver, have lower male rates than the largest cities of eastern Canada. However, these cities have higher rates than most of the rural prairie and Western provinces.

In comparison to total male cancer mortality rates, female rates were less distinct. The five divisions with the highest rates are Montreal, Quebec, Chambly, Ile-Jesus, and Winnipeg.

The highest rates of male lung cancer were in metropolitan Toronto, greater Vancouver, metropolitan Montreal, Hamilton-Wentworth, Hull, Chambly, and Quebec City. All are urbanized areas. Female rates were less distinct. Montreal and the Vancouver metropolitan area had among the highest rates.

The provinces of Quebec, Nova Scotia, and Newfoundland had the highest male large intestine cancer mortality rates. Western Canada manifested the lowest rates. Among the large cities, Montreal and Halifax (Nova Scotia) had among the highest rates. The geographical distribution of female rates resembled that of male rates, but was not as distinct.

High female breast cancer mortality rates were observed in the Montreal region and in less urban areas of eastern Quebec and Nova Scotia. The highest male bladder rates were observed in scattered, urban areas along an axis from Montreal in the northeast to Windsor at the western end of Lake Erie.

The maps did not suggest a consistent association between urbanization and other types of cancer. Other associations were suggested by an analysis of the rates. Specifically, to supplement the maps, a simple quantitative indicator was developed. Approximately 70 census divisions with a population density exceeding 25 persons per square kilometer were identified from a population density map of Canada. Using these as the population of densely settled places, it was hypothesized that if urbanization and cancer mortality rates in Canada were associated, then the clear majority of these census divisions should have higher cancer mortality rates than Canada as a whole.

The expected was usually found. Over two-thirds of these most densely settled census divisions had intestine cancer mortality rates that were higher than the nation's. Between 61 and 64 percent of these census areas had higher than national male bladder and female breast cancer rates, and 58 percent had higher than national male total cancer mortality rates, with the analogous figure for females being 56 percent.

In contrast to the maps, which highlighted the presence of high lung cancer mortality rates in large urban centers, the quantitative analysis suggested that many of the less densely settled urban areas did not have high lung cancer mortality rates. Only 53 percent of the most densely populated census divisions had male lung cancer mortality rates higher than the nation as a whole. The proportion for females was even less, only 45 percent.

As expected, less than one-half of the census divisions had above national rates of most of the other types of cancer. Less than 40 percent of the most densely settled census districts had female tongue, mouth, and pharynx, female pancreas, and female ovary cancer mortality rates higher than the national average. Between 40 and 49 percent of the districts had higher than national cancer mortality rates of male tongue, mouth, and pharynx cancer, male and female rectal cancer, male pancreatic cancer, female lymphatic cancer and female leukemia, and male and female stomach cancer.

Between 51 and 55 percent of the census districts had male lymphatic cancer and male leukemia mortality rates higher than all of Canada. Although unexpected, relatively small urban/rural differences were also found in the United States, and male and female lymphatic rates were also found to be positively associated with urbanization in the 52 nation analysis.

There were two very unexpected results. Almost 60 percent of the most densely settled census districts in Canada had uterine cancer mortality rates higher than the nation, and about three-fourths of the districts had higher prostate cancer mortality rates than the nation. Notwithstanding the possibility that population density may not be the best indicator of urbanization, the uterine and prostate findings were unexpected because the literature reports an inconsistent association with urbanization (Hutchinson, 1976; Christine et al., 1972; Ericson et al., 1976; Pridam and Lilienfeld, 1971; Blair and Fraumeni, 1978).

In summary, Canada demonstrates the expected urban/rural differences for all sites, large intestine, lung, female breast, and male bladder cancer mortality rates. In addition, relatively high male leukemia, male lymphatic and prostate, and uterine cancer mortality rates also seem to be found in densely populated areas of Canada.

THE UNITED KINGDOM

The United Kingdom is an important nation to review for urban/rural differences. In 1973, England and Wales, Northern Ireland, and Scotland ranked ninth, eighth, and third, respectively among the 52 nations in total male and eleventh, ninth, and fourth in total female cancer mortality rates. Particularly high rates of trachea, bronchus, and lung, intestine and rectum, female breast and esophageal cancer characterize the United Kingdom, as do relatively low death rates from stomach and uterine cancer.

G. Melvin Howe (1963, 1970) produced two volumes entitled *National Atlas of Disease Mortality in the United Kingdom*. The atlases cover 1954-58 and 1959-63. Maps of trachea, bronchus, and lung; female breast; uterine; and stomach cancer mortality are provided for more than 150 areas. The maps use the standardized mortality ratio (SMR), a statistic that shows excess or deficit in cancer mortality cases in comparison to the entire nation. The SMR is computed by dividing the number of deaths in the local area by the number that would have occurred if the death rates for each age and sex group had been identical to those for the entire nation. The result is expressed as a percentage. In addition, Howe provides a useful map of population density and urbanization for 1956.

Given the fact that the United Kingdom has among the highest trachea, bronchus, and lung cancer mortality rates among the 52 nations and that the "urban factor" in lung cancer has been sometimes called the "British factor," Howe's observation about the relationship between lung cancer and urbanization is important:

> there appears to be a regular gradation of declining mortality from Greater London, through the provincial conurbations, large, middle-sized and small towns down to rural districts (1963, p. 27).

The SMR's above 140 are found in greater London, Birmingham, Manchester, Lancashire, Liverpool, Leeds, Sheffield, Edinburgh, Glasgow, Dundee, Aberdeen, and other urban areas. The male and female maps for lung cancer are very similar.

The female breast cancer maps are not distinctive, since the relationship between urbanization and the distribution of the disease is inconsistent. There are high mortality areas in the greater London area, but other parts of London have SMR's below 100. Much of Scotland has high rates; Northern Ireland has low rates. Northern England has relatively low rates, whereas southern England has relatively high rates.

Cancer of the uterus does not suggest an urban excess. The highest SMR's are found in rural, northern and western England and Wales and eastern and southern Scotland.

The last of the four types of cancer presented in Howe's atlases is stomach cancer. In the first atlas he states that "the ratios for stomach cancer in the urban districts of practically every county in the United

Kingdom are higher than those for the corresponding rural districts" (1963, p. 28). But in the second atlas, Howe (1970, p. 106) concludes that "no particular urban or rural association is suggested by the maps." Indeed, the most prominent region for stomach cancer is Wales, whereas greater London has among the lowest rates, the converse of the lung cancer maps.

Overall, the Howe atlases point to an obvious urban/rural gradient in lung cancer. However, there is little evidence to suggest a similar gradient for uterine, stomach, and female breast cancer mortality rates.

THE CHANGING RELATIONSHIP BETWEEN URBANIZATION AND RISK FACTORS IN THE UNITED STATES – AN INITIAL LOOK

Cancer mortality differences between aggregates of urban and rural areas of the United States have been decreasing. Before determining whether successively more urbanized areas also demonstrate narrowing of urban/rural cancer gradients, it is important to assess whether the narrowing of the urban/rural gradient may be due to changing geographical relationships between urbanization and factors thought to increase or decrease cancer risk.

Data shortcomings necessitate that an initial examination be undertaken and the results viewed with caution. The available temporal data are at the state scale and usually begin about 1950. This is a gross scale and the time span is not long. Notwithstanding these limitations, the data are adequate to determine whether risk factors have been and continue to be concentrated in urban areas of the United States. Whereas the state data are useful for describing trends, the idea of constructing a multivariate statistical model of change in cancer rates as a function of change in risk factors was rejected. The reasons for this choice were the uneven quality of the data, the serious problem of multicollinearity among the risk factors, and the exacerbated problem of ecological fallacy with state scale data (Cleek, 1979; also see Chapter 2 for a discussion).

State data were available for cigarette sales, alcohol consumption, socioeconomic status, and the presence of white foreign-born and production workers. Completely absent are data for environmental contamination, sunlight, man-made radiation, congenital factors and drug use. According to John Higginson (1979), the founding director of the World Health Organization's International Agency for Cancer Research, these missing factors, along with unknown factors, account for 25 to 30 percent of cancers (see Wynder and Gori, 1977 and Doll and Peto, 1981 for similar estimates). Although there are no time series for diet and life-style, the presence of foreign-born whites and the socioeconomic status are used as major predictors of diet and life-style insofar as cancer risk is concerned.

Cigarette Smoking and Urbanization

With a few exceptions (Burch, 1976; Eysenck and Eaves, 1980), most researchers consider cigarette smoking to be the most important defined environmental factor contributing to cancer. Higginson (1979) attributes 30 percent of male cancers and 7 percent of female cancers to tobacco. Most of the literature falls within ± 10 percent of his estimate for males and within 3 percent for females.

Have an excess of long-term, deep-inhaling, heavy smokers of high tar and nicotine tobacco products lived in urban areas? The only systematic historical data are annual state totals of per capita cigarettes sold (Tobacco Tax Council, 1980). These data fall far short of the desired data (Greenberg et al., 1979). For example, they do not include cigarettes smuggled into states to avoid taxes. Nevertheless, they can be used to follow the changing relationship between cigarette smoking and urbanization in the United States.

Among the major factors influencing per capita cigarette sales at the state scale are the following: socioeconomic status, tax rate, tourism, age structure of the population, state price differentials, ethnicity, religion, and region (the Northeast is high). The Advisory Commission on Intergovernmental Relations (1977) used the above characteristics to predict state per capita cigarette sales. The resulting r^2 of .95 clearly indicates that they were successful and that any relationship between urbanization and cigarette smoking must control for very high or very low taxes and price differentials and tourism. When combined with the absence of data in 1950 for some states, and after the elimination of states with very high and very low taxes and prices, only 35 states were selected for the analysis.

The 35 states were ranked according to their urbanization in 1950, 1960, and 1970 and their per capita cigarette sales. Spearman rank correlations were made between the two ranked data sets. In 1950, the rank correlation between percentage urban and per capita cigarette sales was .64 (significant at .001). The four most urban states among the 35 (New Jersey, New York, Rhode Island, and Connecticut) ranked second, third, sixth, and seventh in per capita cigarette sales.

Per capita cigarette sales in the United States increased 6 percent between 1950 and 1970. The rank correlation between urbanization in 1970 and per capita cigarette sales in 1970 dropped to .43 (significant at .01). The same four most urbanized states ranked second, seventh, tenth, and twelfth in per capita cigarette sales.

The most recent per capita cigarette sales data are for 1980. Rank in urbanization in 1970 was used for the correlation because 1980 census data were unavailable when the calculations were made. The spatial distribution of urbanization in the United States has not substantially changed, as indicated by the rank correlation of .99 between percent urban in 1970 and 1960 and .93 between percent urban in 1970 and 1950. The rank correlation between urbanization in 1970 and per capita cigarette smoking in 1980 was .11 (not significant at .05). Of the four most urban states, only Rhode Island was among the top ten of the 35 states in per capita cigarette sales in 1980. After controlling for bootlegging, the Advisory Commission on Intergovernmental Relations also found no relationship between state cigarette sales and population density in 1975.

In short, although the limitations of the data were recognized, a moderately strong association was found between urbanization and per capita cigarette smoking in the United States in 1950. This relationship has gradually eroded, and by 1980 it was insignificant. If this conclusion is correct at smaller geographical scales, it would certainly *help* explain the narrowing of the gap between urban and rural areas in lung and probably oral, esophageal, larynx, bladder, and pancreatic cancer mortality rates.

Alcohol Consumption and Urbanization

Higginson (1979) attributes 5 percent of male and 3 percent of female cancers to a combination of excessive alcohol use and smoking. The oral cavity, larynx, and esophagus, and perhaps the liver, are commonly cited as the places at risk (Rothman, 1975). The only available time series data on alcohol consumption in the United States comes with virtually the same limitations as the cigarette smoking data. Gavin-Jobson Inc. (1955–1980) prepare annual totals of distilled spirit consumption. Data were obtained from a representative of the company; alcohol data for 43 states for 1955, 1969, and 1979 were compared to percent urban for 1955 (an average of 1950 and 1960), 1970, and 1970, respectively.

The results parallel those for cigarette sales. The Spearman rank correlation between rank in per capita distilled spirits consumption in 1955 and rank in percentage urban in 1955 was .63 (significant at .001). When the same analysis was repeated with 1969 data, the correlation fell to .52 (significant at .001). And when repeated once more with 1979 data, the rank correlation fell to .35 (significant at .01).

The urbanization of life in the United States and its impact on alcohol consumption has been noted by the editorial staff of the *Liquor Handbook* (Gavin-Jobson, Inc. 1980, p. 66):

> Millions moved to metropolitan areas and took up urban ways of life, characterized for one thing by more liberal attitudes toward use of alcoholic beverages.

Overall, there is increasingly less distinction between cities, urban areas, and the entire United States in both cigarette smoking and alcohol consumption. Along with smoking and diet, the changing geography of alcohol use should help explain the narrowing of the urban/rural gap in cancer mortality rates of the tongue/mouth area, larynx, and esophagus.

The White Foreign-Born, Socioeconomic Status, and Urbanization

Higginson (1979) attributes 30 percent of male and 63 percent of female cancers to life-style. According to him, the most important components of life-style are diet (43% of female cancers, 25% for males) and cultural patterns (20% for females, 5% for males). There are no direct indicators of life-style. However, the presence of foreign-born and socioeconomic status are strong clues to life-style with respect to cancer.

As will be seen below, the white foreign-born have had distinctive cancer profiles. To the extent that they have been concentrated in urban areas and have become an important segment of the urban population, the cancer profiles of the white foreign-born should be far more characteristic of urban than rural areas.

Foreign-born whites have constituted a major, but declining component of the population of the most urbanized states. In 1940, 8.7 percent (11.4 million) of the population of the United States was white foreign-born. Almost 60 percent of these 11.4 million people lived in the seven most urbanized states: New Jersey, New York, Massachusetts, Rhode Island, California, Connecticut, and Illinois. The white foreign-born comprised 17 percent of the population of the seven states compared to only 5 percent of the population of the remainder of the nation.

A decade later, the white foreign-born population decreased from 11.4 to 10.2 million, 6.7 percent of the population of the United States. It constituted 13 percent of the population of the seven heavily urbanized states and 4 percent of the population of the remainder of the United States. Along with their descendants, the white foreign-born were a major component of the population of the most urban states in 1940 and 1950.

Using 1950 data, Haenszel (1961) compared cancer mortality profiles of foreign-born whites with those of native-born whites. The profiles are standard mortality ratios prepared from 34,000 death records in 35 states. The profiles are striking. The foreign-born, white male, total cancer mortality SMR in 1950 was 124; for foreign-born, white females it was 106. The highest SMR's were the following: 145 for males and 128 for female digestive system cancer (greater than 175 for esophageal and stomach cancer of both sexes) and 133 for male and 139 for female tracheal, bronchial, and lung cancer. Lymphoma and leukemia SMR's for the white foreign-born population also exceeded 100, though by only 10 to 15 percent; SMR's for female breast, cervix, uterus, and bladder cancer and for prostate cancer were below 90. Overall, the concentration of the white foreign-born in the most urbanized states in 1940 and 1950 implied higher rates for some types of digestive system and lung cancer and lower rates for reproductive system cancers in the most urban states.

Two decades later, in 1970, foreign-born whites were much less prominent. They constituted less than 5 percent of the population of the United States and 9 percent of the population of the seven most urbanized states. It follows that their effect on life-style through diet and culture should be far less than in 1940 and 1950. The gradual disappearance of the white foreign-born should be manifested in a reduction of the difference between urban and rural areas in digestive system, lung, and reproductive organ cancers.

High socioeconomic status is associated with high rates of female breast cancer and poverty with high rates of stomach, liver, lung, and uterine cancer. The 1950, 1960, and 1970 censuses were used to measure the number of poor and wealthy families in the most urbanized states. Given the limited income ranges of the 1950 census, the 1960 and 1970 censuses were used to measure relative poverty. The 1950 and 1970 censuses were used to measure relative affluence. It should be noted that these data could not be adjusted for regional differences in the cost of living because of data deficiencies. The extent of poverty is probably overstated and the degree of affluence understated in some of the rural states and vice versa in the urban states, especially in 1950.

The censuses show that although the most urban states have many more affluent families than they do poor people, in comparison to the most rural states, the gap between the urban and rural states is narrowing. In 1959, the median family income of the United States was $5,660, but more than 21 percent of the families had an income of $3,000 or less (53 percent of the national median). The average of the seven most urban states was only 13.3 percent compared to 34 percent for the eight least urbanized states (Mississippi, North Carolina, North Dakota, South Dakota, South Carolina, Vermont, Virginia, and West Virginia).

A narrowing of the gap is apparent by 1970. The 1969 median family income of the United States was $9,586, but more than 20 percent of American families had an income less than $5,000 (52 percent of the national median). The eight most rural states continued to have the largest percentage of poverty families at 50 percent or less of the national average. However, the average of the eight least urban states declined from 34 percent poverty families in 1959, to 28.6 percent in 1969. Although the proportion of poverty families declined in the eight most rural states, it increased in the seven most urban states from 13 percent in 1959 to 14.7 percent in 1969. Thus, there is a trend, albeit not a sharp one, to a weaker association between very low socioeconomic status and rural areas. This change should be manifested in a narrowing of the urban/rural gap for those diseases associated with poverty.

The difference between the most urban and the most rural states in the number of affluent families remained stable between 1949 and 1969. In 1949, the median family income of the United States was $3,073, with 20 percent of the families earning at least $5,000 (1.63 times the national median family income). Twenty-four percent of the families in the seven most urban states and 13 percent of those in the eight most rural states earned at least $5,000. Two decades later, 20 percent of the families had an income of $15,000 (1.56 times the national median). The proportion of families in the seven most urban states earning at least 1.56 times the national median shrank from 24 percent in 1949 to 23 percent in 1969. In the eight least urban states, it went from 13 to 12.5 percent.

In summary, although life-style and diet cannot be directly measured by available data, there has been an important narrowing of the gap between the most urban and most rural states in two factors that bear upon life-style and diet. The decline of white foreign-born and a reduction in the association between poverty and rural location should be manifested in decreasing differences between urban and rural areas in especially stomach, esophagus, liver, lung, and uterine cancers.

Manufacturing and Urbanization

The diagnosis of occupational cancers began more than two centuries ago, yet the role of the workplace in carcinogenesis is still hotly debated. Higginson (1979) assigns 6 percent of male cancers and 2 percent of female cancers to occupation. Other estimates range from as low as 1 percent to more than 20 percent for males. Production workers are generally considered the largest segment of working population at high risk. Epidemiological studies have drawn attention to asbestos, chemical, petroleum, and many other specific compounds, and counties with these industries have been found to manifest excess cancers (e.g., Blot et al., 1979).

Production jobs are not equally hazardous. Accordingly, it would be helpful to have risk coefficients by industry group. In the absence of such coefficients, a ratio of production workers to residents was used as a simple indicator of potential exposure.

The geographical distribution of production workers in the United States has gradually changed. In 1939, when a good deal of production capacity was idle, 7.8 percent of the population of the seven most urbanized states were production workers, compared to 4.9 percent in the eight least urbanized states. In 1947, 9.8 percent of the population of the seven urban states were production workers, compared to 6.1 in the eight rural states. Two decades later, the number of production workers in the seven most urban states rose 6 percent, but fell to 7.4 percent of the total population. The number of production jobs rose 43 percent in the eight least urban states and rose to the same 7.4 percent of the population as in the seven urban states. By 1972, the number of production jobs in the seven most urban states had fallen below 1947 levels and was 6.6 percent of the population. Production jobs continued to increase in the eight more rural states and comprised 7.7 percent of the population. Overall, production jobs and their associated risks are no longer confined to the most urban states.

As expected from the above presentation, differences in cancer mortality rates between the most and the least urban states have narrowed. During 1950–54, the white male total cancer mortality rate of the seven most urban states was 138 percent of the rate of the eight most rural states (Table 4–7). By the close of the 1960's, the difference had decreased from 38 to 16 percent. The change was much less precipitous for white females, decreasing from 23 percent during 1950–54 to 19 percent during 1965–69.

Table 4-7. Changing Cancer Mortality Rates in the Seven Most Urban and Eight
Least Urban States in the United States, 1950-54 and 1965-69

Area	Population	Rates/100,000		
		(1) 1950-54	(2) 1965-69	$\frac{(2)}{(1)} \times 100$
(A) Seven most urban states	white male	183.8	195.1	106
(B) Eight most rural states	white male	133.3	168.0	126
(C) Seven most urban states	white female	148.9	135.2	91
(D) Eight most rural states	white female	121.4	113.2	93
(A/B)		138	116	
(C/D)		123	119	

Other Factors: Data, Disease Competition, and Population Migration

Any suggestion that the narrowing of urban/rural cancer gradients was
due to changes in the spatial distribution of risk factors must try to take
into account or at least mention differences in medical diagnosis and
recording practices, other causes of death, and, especially, population
migration. There is no literature that presents the changing accuracy of
medical diagnosis and recording at the state scale over time. Although
mortality data do exist for other causes of death, an adequate time series
does not exist at the state scale.

Among the other factors, only the role of interstate migration can be
examined. Ignoring vacations, brief trips, and short-term military service,
the greater the percentage of persons living in a state who were born and
who had always resided in that state, the less the role of interstate migration as a factor distorting state cancer profiles. The closest approximation
to the desired data are census records of native-born persons who were
born in and continued to reside in the same state. To be conservative, it
is assumed that foreign-born persons could have been exposed to environments that differed from those of natives of the state. Census data also
indicate the origin of persons born out of state and sources of interstate
migration. Overall, if the vast majority of the residents of the most rural
states were born in these rural states, or were born in other rural states
with low cancer mortality rates, and rarely migrated to the large urban
states, then interstate migration is not an important factor in accounting
for changes in cancer mortality in the group of states that have manifested
among the highest increases in white male cancer mortality rates.

In 1970, more than 95 percent of the American population was native born. In the eight rural states, the percentage of native-born Americans ranged from 95 to more than 99 percent. Seventy-three percent of the population of the eight rural states was both born and resided in their respective states in 1970. Four of the states had a large foreign-born population or a population born in other states: North Dakota, South Dakota, Virginia, and Vermont. Accordingly, migration cannot be dismissed a priori as a factor for four of the eight states. The other four states (Mississippi, North Carolina, South Carolina, and West Virginia)

Table 4–8. Place of Birth of Persons Residing in Mississippi, North Carolina, South Carolina, and West Virginia in 1970

1. State	Born and Residing in State (1970)	State	Born and Residing in State (1970)
Mississippi	81%	North Carolina	79%
2. State	Born in Following States and Residing in Mississippi (1970)	State	Born in Following States and Residing in North Carolina (1970)
1. Alabama		Alabama	
2. Arkansas	10%	Florida	9%
3. Georgia		Georgia	
4. Kentucky		Kentucky	
5. Louisiana		South Carolina	
6. North Carolina		Tennessee	
7. Oklahoma		Virginia	
8. Tennessee		West Virginia	
9. Texas			
Total 3. 1 + 2	91%	Total	88%
4. Born in Seven Most Urban States and Residing in Mississippi (1970)	1.5%	Born in Seven Most Urban States and Residing in North Carolina (1970)	2.5%

had in-state birth rates ranging form 76 to 81 percent. Moreover, determination of places of birth of those who resided in these four rural states in 1970, but who were born out of state, showed that most of the people were born in neighboring rural states and few people in urban states. Overall, in these four states, an average of 79 percent of the population was born in the state of residence, and an average of 10 percent was born in neighboring states. Only an average of 2 percent was born in the seven most urban states (Table 4–8).

State	Born and Residing in State (1970)	State	Born and Residing in State (1970)
South Carolina	76%	West Virginia	81%

State	Born in Following States and Residing in South Carolina (1970)	State	Born in Following States and Residing in West Virginia (1970)
Alabama		Kentucky	
Florida	11%	Maryland	11%
Georgia		Ohio	
Kentucky		Pennsylvania	
North Carolina		Virginia	
Tennessee			
Texas			
Virginia			
Total	87%	Total	92%

Born in Seven Most Urban States and Residing in South Carolina (1970)		Born in Seven Most Urban States and Residing in West Virginia (1970)	
2.7%		1.4%	

The latter four states have exhibited high increases in cancer mortality (Table 4–9). These rates of increase were higher during 1950–75 than the four most rural states for which migration could not be ruled out as a factor. White male cancer mortality rates in Mississippi, South Carolina, and West Virginia were about 83 percent of the national rate during 1950–54. A decade later, the difference had decreased to 10 percent. By 1970–75, the white male rates of three of the four rural states were the same as those in the far more urbanized United States. North Carolina had an even lower white male cancer mortality rate than the other three rural states when the study period began; but two decades later, the white male rate in North Carolina was 93 percent of the national rate.

Could population migration have played an important role in these states? A comparison was made of white male cancer mortality rates in the four rural states, their neighboring states, and the seven most urban states. Weighted average cancer mortality rates were calculated for each of the four rural states. The weighted average consists of the population born and residing in the rural state (e.g., Mississippi), multiplied by the cancer mortality rate in that state. Added to the first total is the weighted average of the number of persons born in the neighboring states and residing in the rural state in 1970 times the cancer mortality rates of the neighboring states. For example, the weighted average, white male cancer mortality rate of the nine neighboring states of Mississippi is only 1.9 percent higher than Mississippi's white male cancer mortality rate. Accordingly, even if the conservative assumption is made that persons born in the neighboring states, but living in Mississippi, have the cancer mortality rates of the neighboring states, the impact on Mississippi's

Table 4–9. Changing White Male Cancer Mortality Rates in Mississippi, North Carolina, South Carolina, and West Virginia, 1950–75

	Rate/100,000					
State	1950–54	1955–59	1960–64	1965–69	1970–75	$\dfrac{\text{Rate } 1970\text{–}75}{1950\text{–}54} \times 100$
(A) Mississippi	138.4	142.5	163.0	176.6	189.0	137
(B) North Carolina	117.7	132.6	145.7	156.5	176.6	150
(C) South Carolina	135.4	149.7	153.1	174.4	190.0	140
(D) West Virginia	132.3	145.5	161.3	176.4	188.8	143
(A)/U.S. rate × 100	85	84	94	97	100	—
(B)/U.S. rate × 100	72	78	84	86	93	—
(C)/U.S. rate × 100	83	88	88	96	100	—
(D)/U.S. rate × 100	81	86	93	97	100	—

white male rate should be to increase it less than 1/100,000. The seven most urban states have a much higher cancer mortality rate than Mississippi. However, since persons born in the seven urban states and residing in Mississippi were only 1.5 percent of the Mississippi population in 1970, their impact on the Mississippi white male cancer mortality rate would again be less than 1/100,000.

With respect to North Carolina, if the same conservative assumptions are made as were made for Mississippi, migrants from its neighboring states would raise its white male rates by 1 to 1.5/100,000, and migrants from the seven most urban states would also increase the white male rate between 1 and 1.5/100,000. The analogous impact on South Carolina would be a decrease of less than 1/100,000 from its neighbors and an increase of 1 to 1.5/100,000 due to migrants from the seven most urban states. Finally, in the case of West Virginia, the neighboring states would increase the rate about 2/100,000 and the seven most urban states would increase it by less than 1/100,000. In short, the place of residence by place of birth data suggest a minimal impact on cancer mortality rates of the four rural states by interstate migration.

It can be argued that place of residence by place of birth data do not take repeated interstate migration into account. The worst scenario would involve persons moving from a low risk environment to a high risk environment out of the state only to return to the low risk environment in their later years. The data do not support the worst scenario. A review of migration data for 1955–60, 1965–70, 1970–73, and 1978–79 suggest that about 95 percent of the people who reside in the four rural states migrate within the state of residence or one of the neighboring states included in the previous analyses. Only about 1 percent of the residents of the four states came from the seven most urban states. Although these census and Internal Revenue Service (Engels and Healy, 1980) data are admittedly imperfect, they give little credence to interstate migration as a factor in the case of these four rural states.

Additional analyses were made of migration and place of birth data of the white male population and of age groups. In all four states, the white male population had a higher probability of being born out of state than the other populations. However, the difference is small except for the 20- to 24-year-old age group, which corresponds to World War II period births. In conclusion, unless some very conservative assumptions about migration from the urban and neighboring states to the rural states are made, one is left with the conclusion that the substantial increases in white male cancer mortality rates in the four rural states were not strongly influenced by the interstate migration of people from states in which the risk from cancer was considerably higher.

Two extremely conservative assumptions about the nature of interstate migration could change the conclusion. First, since not every year between 1950 and 1975 is covered by available migration data, it is possible that the migration data do not accurately reflect origin in and destinations to the four rural states. This possibility is highly unlikely because of the consistency in the patterns of interstate migration flows (Engels and Healy, 1980). Second, perhaps all the migrants to the four states have come from the highest cancer rates areas of the neighboring states. For example, if one assumes that the migrants to the four states from the neighboring states came from areas in the neighboring states with cancer mortality rates 50 percent higher than the state as a whole, then interstate migration becomes a significant explanatory factor. For example, if the 50 percent higher assumption is applied to the nine states that are Mississippi's source region and the seven most urban states, Mississippi's white male rate would increase 11/100,000, not less than 1/100,000. The same assumption applied to North Carolina would have resulted in an increase of the white male rate by 14/100,000, not 2 to 3/100,000. With respect to South Carolina, the extremely conservative most urban county origin hypothesis would have increased the white male rate by 13/100,000, not 1/100,000. The analogous change for West Virginia, would be 14/100,000, not 2.5/100,000. Although possible, it seems improbable that all migrants to Mississippi, North Carolina, South Carolina, and West Virginia from Georgia come from Atlanta, all migrants from Alabama from Birmingham, all migrants from New Jersey from Newark and Jersey City, and so forth.

In summary, interstate migration of high risk persons may help to explain changes in cancer mortality rates in states like Alaska, Arizona, Florida, Hawaii, and Nevada and, to a lesser extent, California, Colorado, and New Mexico. Interstate migration does not appear to be an important factor in Mississippi, North Carolina, South Carolina, and West Virginia, and perhaps other states that exhibited pronounced increases in cancer mortality rates. In the absence of data, competing causes of death and differences in medical practices between urban and rural areas cannot be ruled out.

The changing distribution of etiological factors seems to play an important role because notwithstanding the caveats about the data, it is clear that when the study period began, many measurable indicators of increased cancer risk were far more characteristic of the most urban states than they were of the least urban states. By the late 1960's and early 1970's, the sharp distinction between the most and the least urban states with respect to general indicators of risk were blurred. Although only suggestive because of the primitive data that were used and the state scale of analysis, the empirical findings about the risk factors are consistent with the observation that spatial convergence of age-, race-, and sex-controlled cancer mortality rates has occurred.

SUMMARY

This chapter began with three broad questions, the answers to which may be summarized as follows:

1. How do the actual urban/rural differences in cancer in the United States compare to differences that are suggested in the literature?

During 1950–54, the expected and actual profiles conformed closely. The most urban counties had white population cancer rates at least 40 percent higher than the least urban counties for most of the expected types of cancer: male bladder, male kidney, male larynx, male lung, male esophagus, male and female rectum, male tongue/mouth, and male large intestine. Female breast cancer and the other expected high urban/rural ratios for females exceeded 1.3. By 1970–75, all but one of the expected high urban/rural ratios (male tongue/mouth) declined to less than 1.4. Overall, the strongly urban/rural ratios for total cancer decreased from 1.35 to 1.12 for white males and from 1.16 to 1.12 for white females. With one exception, the only widening of urban/rural differences was for declining mortality rates for such cancers as stomach and liver. The exception is white female respiratory cancer (trachea, bronchus, lung, larynx), which exhibited a more pronounced increase in urban than in rural areas due to marked increases in rates of white females 55 years of age and older. Marked but declining urban/rural differences were also identified for nonwhites.

2. Is there an association between urbanization and cancer mortality rates in other nations?

Using 52 nations and 1973 as a base line, an association was found between urbanization and the following types of cancer: male and female intestine; female breast; male and female other lymphatic; male and female lung; male and female esophagus; male rectum; and male and female all types. Female other uterine cancer mortality rates were observed to decrease as urbanization increased. These findings again conform very closely to expectations derived from the literature. Urban/rural differences were also found within Poland, Norway, the United Kingdom and Canada.

3. How has the association between urbanization and cancer risk factors changed in the United States?

During the early 1950's, the highest per capita rates of cigarette and alcohol consumption were found in the most urban states. These states also had a large percentage of European-born, white Americans, relatively high socioeconomic status in comparison to the rural states, and a large concentration of production workers. The combination of these attributes in the most urban states suggests high cancer risk for many types of cancer that exhibited pronounced urban/rural differences including digestive, respiratory, and urinary systems and female breast.

Two decades later, nearly all these characteristics were no longer urban American characteristics. They were American characteristics. It is hypothesized that this reduction of differences in the geography of risk factors has led to the spatial convergence of cancer rates in the United States. Changing medical practices, disease competition, and migration may also play a role. Of the three, only interstate migration could be examined as a factor. In four of the most rural states that have shown among the highest increases in cancer mortality, interstate migration played an unimportant role in the increase.

REFERENCES

Adelstein, A. 1972. "Occupational mortality: Cancer," *Ann. Occup. Hyg.* 15: 53–57.

Advisory Commission on Intergovernmental Relations 1977. *Cigarette Bootlegging: A State and Federal Responsibility*, Washington, D.C.: The Commission.

Anderson, R., Ishida, K., Ishimaru, T., and Nishiyama, H. 1970. "Geographic aspects of malignant lymphoma and multiple myeloma," *Amer. J. Pathol.* 61: 85–97.

Blair, A., and Fraumeni, J. 1978. "Geographic patterns of prostate cancer in the United States." *J. Natl. Can. Inst.* 61: 1379–1384.

Blair, A., 1979. "Leukemia among Nebraska farmers: A death certificate study," *Amer. J. Epidemiol.* 110: 264–273.

Blot, W. J., Stone, B. J., Fraumeni, J. F., and Morris, L. E. 1979. "Cancer mortality in U.S. counties with shipyard industries during World War II." *Environ. Res.* 18 (2): 281–290.

Burch, P. 1976. *The Biology of Cancer.* Baltimore: University Park Press.

Bureau of Epidemiology, Health and Welfare, Canada 1980. *Mortality Atlas of Canada,* Vol. 1, Hull, Quebec: Canadian Government Publishing Center.

Burmeister, L. 1980. unpublished.

Christine, B., Groff, W., Pitt, T., and Chapple, M. 1972. "The relationship of incidence of cervical cancer and socioeconomic status in seven cities, 1959-1964." *Conn. Med.* 36: 80-83.

Cleek, R. 1979. "Cancers and the environment: The effect of scale." *Soc. Sci. Med.* 13D: 241-247.

Doll, R., and Peto, R., 1981. *The Causes of Cancer.* New York: Oxford University Press.

Donham, K., Berg, J., and Sawin, R. 1980. "Epidemiologic relationships of the bovine population and human leukemia." *Amer. J. Epidemiol.* 112: 80-92.

Engels, R., and Healy, M. 1980. "Measuring Interstate Migration Flows: An Origin-Destination Network through International Revenue Service Records." Paper presented at annual meeting of the Northeast Regional Science Association, Boston, Mass., May 3, 1980.

Ericson, J. L., Karnstrom, L., Mattson, B., and Willgran, J. 1976. "The incidence of breast and cervix cancer in the Swedish population." In H. Bostrom and T. Larsson, eds. *Health Control in the Detection of Cancer,* Stockholm: Almquvist & Wiksell International.

Eysenck, H., and Eaves, L. 1980. *The Causes and Effects of Smoking,* Beverly Hills, Calif.: Sage.

Gavin-Jobson, Inc. 1955-1980. *The Liquor Handbook,* various volumes, New York: Gavin-Jobson, Inc.

Greenberg, M., and Caruana, J. 1979. *Cancer Mortality Patterns in the New Jersey-New York-Philadelphia Metropolitan Regions, 1950-1972,* Part II, Trenton, N.J.: Department of Environmental Protection.

Haenszel, W., 1961. "Cancer mortality among the foreign born in the United States." *J. Nat. Can. Inst.* 26: 37-132.

Higginson, J. 1979. "Cancer and the environment: Higginson speaks out." *Science* 205: 1363-1365.

Howe, G. M. 1963. *National Atlas of Disease Mortality in the United Kingdom,* London: Nelson and Sons.

Howe, G. M. 1970. *National Atlas of Disease Mortality in the United Kingdom,* London: Nelson and Sons.

Hutchinson, G. B. 1976. "Epidemiology of prostatic cancer." *Semin. Oncol.* 3: 151-159.

Lilienfeld, A., Levin, M., and Kessler, I. 1972. *Cancer in the United States.* Cambridge, Mass.: Harvard University Press.

Mancuso, T. 1977. "Lung cancer among black migrants," *J. Occup. Med.* 19: 531-532.

Mancuso, T., and Sterling, T. 1975. "Lung cancer among black and white migrants in U.S." *J. Natl. Med. Assoc.* 67: 106-112.

Mason, T., McKay, F., Hoover, R., Blot, W., and Fraumeni, J. 1976. *Atlas of Cancer Mortality among U.S. Nonwhites: 1950-1969.* Washington, D.C.: DHEW Pub. No. NIH 76-1204.

Nishiyama, H., and Inove, T. 1970. "Some epidemiological features of Hodgkin's disease in Japan." *Gann* 61: 197-205.

Pridan, H., and Lilienfeld, A. 1971. "Carcinoma of the cervix in Jewish Women in Israel, 1960-1967: An epidemiological study." *Isr. J. Med. Sci* 7: 1465-1470.

Rothman, K. 1975. "Alcohol." In J. F. Fraumeni, Jr., ed. *Persons at High Risk of Cancer.* New York: Academic Press, pp. 139–150.

Segi, M. 1978. *Age-Adjusted Death Rates for Cancer for Selected Sites (A Classification) in 52 Countries in 1973.* Nagoya, Japan: Segi Institute.

Straszewski, J. 1976. *Epidemiology of Cancer of Selected Sites in Poland and Polish Migrants,* Cambridge, Mass: Ballinger.

Tobacco Tax Council, Inc. 1980. *The Tax Burden on Tobacco,* Vol. 15, Richmond, Va.: Tobacco Tax Council.

Todd, C. 1981. Intraracial Differences among Blacks on Perceived Health Status, draft of Ph.D thesis, Columbia University.

United Nations, Department of International Economic and Social Affairs 1980. *Patterns of Urban and Rural Population Growth,* Population studies, No. 68, New York: United Nations.

U.S. Bureau of the Census, 1973. *County and City Data Book,* Washington, D.C.: U.S. Government Printing Office.

U.S. Bureau of the Census, 1958, 1964, 1970, 1976. *Statistical Abstract of the United States,* Washington, D.C.: U.S. Government Printing Office.

U.S. Bureau of the Census, 1952, 1956, 1967, 1972. *County and City Data Book,* Washington, D.C.: U.S. Government Printing Office.

U.S. Bureau of the Census, 1972. *Census of the Population: 1970, General, Social and Economic Characteristics,* Final Report PC(1)-C1 United States Summary, Washington, D.C.: U.S. Government Printing Office.

U.S. Bureau of the Census, 1972. *U.S. Census of Population 1960. Subject Reports. Mobility for States and State Economic Areas,* Final Report PC(2)-2B, Washington, D.C.: U.S. Government Printing Office.

U.S. Bureau of the Census, 1973. *Census of Population: 1970 Subject Reports, Final Report PC(2)-2A, State of Birth* Washington, D.C.: U.S. Government Printing Office.

Waterhouse, J., Muir, E., Correa, P., and Powell, J., eds. 1976. *Cancer Incidence in Five Continents,* Vol. III. Lyon: IARC Scientific Publications, No. 15.

Wynder, E., and Gori, G. 1977. "Contribution of the environment to cancer incidence: An epidemiolgic exercise." *J. Natl. Can. Inst.* 58: 825–832.

5 The Metropolises: Child, Teenage, and Total Population Cancer Mortality Profiles

This chapter will closely examine cancer mortality patterns within the most urbanized areas of the United States. Three major questions are posed:

1. Do the metropolitan areas of the Northeast, Northcentral, South and West differ in cancer mortality profiles?
2. Are there differences between the metropolises of the United States as a whole and the entire nation in child and teenage cancer mortality rates?
3. Within the group of most urbanized areas, are there regional differences in child and teenage cancer mortality profiles?

Before the results are presented, decisions about the scope of the analysis and the presentation will be discussed. Chapter 4 included every county in the continental United States that was at least 75 percent urban in the strongly urban group, and 78 percent of the strongly urban counties were in Standard Metropolitan Statistical Areas (SMSA's) in 1970. However, even if they had all been in metropolitan areas, many are not part of large metropolitan areas. In order to test the hypothesis that the largest metropolises have had the highest cancer rates, the 25 largest metropolitan regions, each with a population of over one million in 1970, were adopted as a working group (Table 5–1). These 25 metropolises (110 counties) had 35 percent of the national population in 1970. Sixty-three percent (69 of 110) of these metropolitan counties are also strongly urban. Thus, this chapter uses both the strongly urban counties and the major metropolises as bases to measure changes within the group of highly urbanized counties.

Table 5–1. The 25 Major Metropolises

1. Atlanta, Ga.	14. Milwaukee, Wisc.
2. Baltimore, Md.	15. Minneapolis/St. Paul, Minn.
3. Boston, Mass.	16. Newark, N.J.
4. Buffalo, N.Y.	17. New York, N.Y.
5. Chicago, Ill.	18. Paterson/Passaic/Clifton, N.J.
6. Cincinnati, Ohio–Ky.–Ind.	19. Philadelphia, Pa.–N.J.
7. Cleveland, Ohio	20. Pittsburgh, Pa.
8. Dallas, Tex.	21. St. Louis, Mo.–Ill.
9. Detroit, Mich.	22. San Diego, Calif.
10. Houston, Tex.	23. San Francisco/Oakland, Calif.
11. Kansas City, Mo.–Kans.	24. Seattle/Everett, Wash.
12. Los Angeles/Long Beach, Calif.	25. Washington, D.C.–Md.–Va.
13. Miami, Fla.	

A second choice was not to present proportional mortality data in this chapter or in Chapter 6. This decision was made because the proportional mortality rates for the major metropolises are virtually identical to those of the strongly urbanized counties. Age-specific and age-adjusted rates are used for the regional comparisons, but the age-specific rates are not used for the metropolitan counties because they are so similar to those of the strongly urban counties.

CANCER PROFILE OF THE 25 MAJOR METROPOLITAN REGIONS

The cancer mortality profile of the white population of the 25 major metropolises closely resembles that of strongly urban counties (Tables A1–33, A1–34). It is a profile characterized by relatively high rates of respiratory, some digestive, and urinary system cancers, female breast cancer, and some systemic cancers including non-Hodgkin's lymphoma and Hodgkin's disease and relatively low rates of cancer of the cervix uteri and corpus uteri and other skin cancer.

Two decades later, the profile more closely resembles that of the rural and strongly urban aggregates and the continental United States. During 1950–54, white male, total cancer mortality rates of the metropolitan counties were 41, 17, and 5 percent higher than those of the rural aggregate, continental United States, and strongly urban aggregate. Two decades later, the differences had narrowed to 14, 6, and 2 percent. For white females, the gap narrowed from 20, 10, and 3 percent to 15, 7, and 2 percent.

CHANGING REGIONAL CANCER MORTALITY PROFILES
WITHIN THE MOST URBAN AREAS

The first section of this chapter, and Chapter 4, demonstrated that differences in cancer mortality between the major metropolises, the strongly urban counties, and the remainder of the United States have substantially narrowed during the period 1950–75. But perhaps the most urban areas are not uniform in their profiles and rates of change. Are there regional differences in urban cancer mortality rates within the strongly urban and metropolitan aggregates? Since the Northeast and Northcentral regions were urbanized and industrialized before other regions of the United States and were the ports of entry and areas of settlement for most of the foreign-born white population, it is hypothesized that the cancer rates of the most urbanized counties of the Northeast and Northcentral regions were higher than those of the South and West. These differences should have decreased as the nation has become more homogeneous. It is further hypothesized that the cancer mortality profiles of every metropolitan region in the Northeast and Northcentral United States should have demonstrated changes in cancer mortality rates that clearly distinguish them from changes observed in the Southern and Western metropolises.

Age-Adjusted Data: Strongly Urban Counties

Most of the expected differences between Northeast (NE), Northcentral (NC), Southern (S), and Western (W) urban areas were found (Figures 5–1 and 5–2; Tables A1–35 through A1–42). This was done by dividing the cancer mortality rates of the urban areas in each region (e.g., NE, NC, S, and W) by the national rate for the urban areas. During 1950–54, the white male and white female total cancer mortality rates for the counties of the Northeast were, respectively, 11 and 8 percent higher than the entire strongly urban aggregate (Tables A1–35 and A1–39). Northeastern strongly urban/aggregate strongly urban ratios greater than 1.15 were observed for male and female rectum, male larynx, male and female large intestine, male esophagus, and female stomach cancer. The lowest ratio, for female cervix uteri cancer, was 0.85.

Northcentral urban areas had rates at about the average for the strongly urban areas. No Northcentral strongly urban/aggregate strongly urban ratios were above 1.15.

Most of the Southern and Western strongly urban/aggregate strongly urban ratios were below 1.0. The total, white male cancer ratios for the

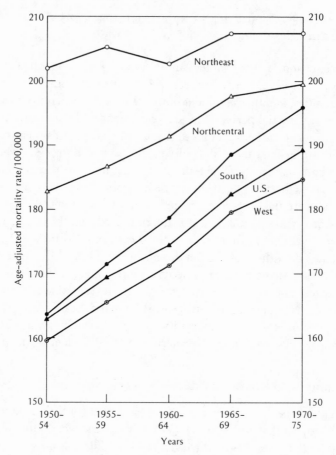

Figure 5-1. Trends in all types cancer mortality, white males: Strongly urban regions, 1950-75

Southern and Western strongly urban aggregates were 0.90 and 0.87 during 1950-54. Only white female melanoma (South and West), cervix uteri cancer (South), and male other skin cancers (South) were above 1.15 during 1950-54. Many cancer ratios were below or were 0.85: female stomach (South and West); male stomach (South); male and female large intestine (South and West); male and female rectum (South and West); male esophagus (South and West); female liver (South and West); male liver (West); male tongue/mouth (West); male larynx (West); male bone (West); female breast (South); ovary (South); and corpus uteri (West).

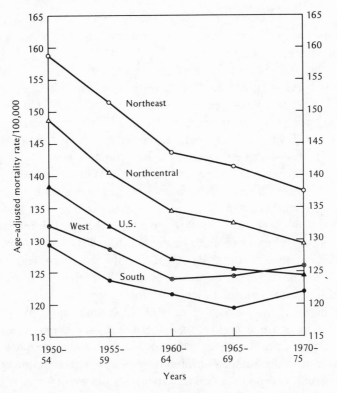

Figure 5–2. Trends in all types cancer mortality, white females: Strongly urban regions, 1950–75

Two decades later, the gap between the strongly urban areas had dramatically narrowed (Figures 5–1 and 5–2). Northeastern strongly urban/aggregate strongly urban ratios for total white male and white female cancers decreased from 1.11 and 1.08 to 1.05 and 1.06. West ratios increased from 0.87 and 0.91 to 0.93 and 0.97. The other two regions fell between these two. Nearly all the ratios for the specific types of cancer were closer to 1.0 during 1970–75 than during 1950–54.

There are some notable exceptions: female lung cancer and female Hodgkin's disease. Beginning with lung cancer, it should be reiterated that female respiratory tract cancer was reported to be the single exception to increasing homogeneity of disease among rural, moderately urban, and strongly urban aggregates (Chapter 4). The strongly urban and moderately urban aggregates had higher rates of increase of cancer of the respiratory system than did the rural counties. Within the set of almost

300 strongly urban counties, counties in the South and the West have clearly exhibited the most pronounced rates of increase of white female respiratory tract cancers.

Data for 1950–54 certainly gave no hints of such changes (Tables A1–39 through A1–42). Among the strongly urban set, the highest white female lung cancer rates were 1.06 (Northeast) and 1.00 (Northcentral). Rates in strongly urban counties in the South and the West were relatively lower (0.92 and 0.90). Indeed, the white female, lung cancer age-adjusted rate in Southern strongly urban counties was only 2 percent higher than the national figure; and the rate in Western counties equaled the national rate. In short, there was no reason to suspect the very rapid increases in Southern and Western strongly urban counties.

By 1960–64, there was clear evidence of a change. Cancer mortality rates in strongly urban counties in the South and the West were higher than the national average. By 1970–75, the Southern and Western strongly urban aggregates were substantially higher than either the Northeast or Northcentral strongly urban aggregates.

Finding that female mortality rates of respiratory cancer in urban and rural aggregates are diverging and that Southern and Western aggregates are diverging from Northeastern and Northcentral aggregates are so unusual that in the author's opinion they clearly deserve followup.

The second exception is female Hodgkin's disease. During 1950–54, the rates varied from 1.5/100,000 in the strongly urban Northeast to 1.2/100,000 in the strongly urban West. The Northeast rate was 15 percent higher and the Western rate was 8 percent lower than the rate for the continental United States. Rates in all four regional aggregates declined during the subsequent score of years. The smallest decrease was in the Northeast (1.5 to 1.4). By 1970–75, the Northeast strongly urban/aggregate strongly urban ratio had increased from 1.07 to 1.27. This means that the difference between the Northeast urban areas and the remaining urban areas in female Hodgkin's disease have increased.

Age-Specific Data: Strongly Urban Counties

Age-group data pinpoint the tendency toward similar cancer profiles among the Northeastern, Northcentral, Southern, and Western strongly urban aggregates (Tables A1–43 through A1–46). During the early 1950's, white male and female age-specific rates in the Northeast, strongly urban counties were higher than rates in the other regional aggregates for every age group 15 years and older. Clear differences were apparent for age groups 55–74. Age-adjusted rates in the strongly urban counties of the Northeast were almost always higher two decades later. Age-specific rates, however, were not consistently higher for males until age group 65 and for females until age group 45. Furthermore, the major differences in age-specific cancer mortality rates between the Northeast and other areas had diminished.

Time Change Indices: Discriminant Analyses of Major Metropolitan Regions

Discriminant analysis (see Appendix 2 for a brief discussion) was used to test the hypothesis that the 25 metropolises of the four regions of the United States exhibited changes in cancer mortality characteristic of their region. The regional groups were taken from U.S. Bureau of the Census definitions (Table 5–2) with one exception. Pittsburgh was placed in the Northcentral region because it is close to the Great Lakes manufacturing cities. As the analysis will show, Baltimore also should have been placed in another region. The Northeast contains six metropolitan areas, the Northcentral region nine, and South six, and the West four. Time change indices between 1960–69 and 1950–59 for 22 types of cancer were used as the input data. Ten-year time periods, the white population, and 22 types of cancer were used because of the author's concern with data reliability. The use of ten-year time periods increases one's confidence in the rates and, in turn, in the time change indices. In addition to using decade rates for only the white population, the types of cancer were limited roughly to those with rates exceeding 5/100,000. The nine types of male and eleven types of female cancer (Table 5–3) accounted for 72 percent of white male and 76 percent of white female cancer deaths during the study period.

The results strongly support the hypothesis that the 25 metropolises do identify strongly with regional patterns of change. The initial results are the regional and grand means for the 22 types of cancer (Table 5–4). Columns 1 through 5 are the time change indices for each of the four regions and the grand means. Adjacent to the means are the regional means divided by the grand means. The strongest deviants from the grand means are underlined. In general, the time change indices are similar for almost every type of cancer; however, 16 of the 88 regional means differ from the grand means by 5 percent or more. They will be the focal points of this discriminant analysis.

As expected, metropolises of the Northeast have the largest number of regional/grand mean comparisons less than 1.00. At the other end of the spectrum, 15 of the 22 regional mean/grand mean ratios for the West

Table 5–2. Regional Location of 25 Most Populous SMSA's

Northeast (NE)	*South (SO)*
1. Boston (BOST)	16. Atlanta (ATL)
2. Buffalo (BUFF)	17. Baltimore (BALT)
3. Newark (NWK)	18. Dallas (DALL)
4. New York (NY)	19. Houston (HOUS)
5. Paterson/Passaic/Clifton (PPC)	20. Miami (MIAM)
6. Philadelphia (PHIL)	21. Washington, D.C. (DC)
Northcentral (NC)	*West (WE)*
7. Chicago (CHIC)	22. Los Angeles (LA)
8. Cincinnati (CINC)	23. San Diego (SD)
9. Cleveland (CLEV)	24. San Francisco (SF)
10. Detroit (DETR)	25. Seattle (SEAT)
11. Kansas City (KC)	
12. Milwaukee (MLW)	
13. Minneapolis (MINN)	
14. Pittsburgh (PITT)	
15. St. Louis (STLO)	

exceed 1.00. Twelve of the 22 types of cancer demonstrate relatively little difference between the four regions: male stomach, male large intestine, male prostate, male bladder, male all types, female large intestine, female rectum, female liver, female pancreas, female breast, female uterus, and female all types. The remaining ten types have at least one strongly deviant regional mean/grand mean ratio.

An initial review of means suggests that male liver, male leukemia, female stomach, female lung, and female cervix are the most discriminating cancers. Briefly, male liver cancer manifested a much less pronounced decrease in the Western metropolises than it did in the North-

Table 5–3. The 22 Types of Cancer Included in the Metropolitan Region Time Change Discriminant Analysis

1. Male stomach (MSTOM)	12. Female intestine (FINT)
2. Male intestine (MINT)	13. Female rectum (FRECT)
3. Male rectum (MRECT)	14. Female liver (FLIV)
4. Male liver (MLIV)	15. Female pancreas (FPANC)
5. Male pancreas (MPANC)	16. Female lung (FLUNG)
6. Male lung (MLUNG)	17. Female breast (FBRST)
7. Male prostate (MPROS)	18. Female cervix (FCERV)
8. Male bladder (MBLAD)	19. Female uterus (FUTS)
9. Male leukemia (MLEUK)	20. Female ovary (FOVAR)
10. Male all types (MALL)	21. Female leukemia (FLEUK)
11. Female stomach (FSTOM)	22. Female all types (FALL)

Table 5-4. Comparison of Regional Means and Grand Means of Time Change Indices for 22 Types of Cancer

Type of Cancer[a]	(1) NE	(1)/(5)[b]	(2) SO	(2)/(5)	(3) NC	(3)/(5)	(4) WE	(4)/(5)	(5) Grand Mean
					Time Change Indices				
1. MSTOM	67.5	102	65.7	99	65.2	98	68.2	103	66.3
2. MINT	102.4	98	106.3	102	104.4	100	106.6	102	104.7
3. MRECT	76.7	94	83.3	102	82.7	101	85.3	104	81.6
4. MLIV	83.9	102	83.3	101	78.1	95	88.1	107	82.4
5. MPANC	111.7	99	105.9	94	117.1	104	114.9	102	112.8
6. MLUNG	136.5	94	144.3	100	149.5	103	148.1	102	144.9
7. MPROS	90.5	97	91.4	97	97.1	104	94.8	101	93.8
8. MBLAD	90.4	97	92.4	99	94.8	102	95.7	103	93.3
9. MLEUK	99.6	95	100.3	96	110.5	106	105.7	101	104.7
10. MALL	100.6	96	106.1	101	105.9	101	107.1	102	104.8
11. FSTOM	63.7	98	72.7	111	60.9	93	66.4	102	65.3
12. FINT	90.9	97	93.7	100	96.7	103	94.0	100	94.1
13. FRECT	75.4	99	76.2	100	77.8	102	73.9	97	76.2
14. FLIV	72.5	102	68.0	96	71.6	101	73.4	103	71.2
15. FPANC	104.9	97	109.3	101	108.2	100	113.2	104	108.5
16. FLUNG	151.5	94	165.4	103	153.5	95	186.9	116	161.2
17. FBRST	98.8	100	101.5	103	98.1	99	95.1	97	98.6
18. FCERV	76.1	105	66.8	92	75.3	104	70.6	97	72.7
19. FUTS	72.0	101	69.2	97	72.5	101	71.7	100	71.5
20. FOVAR	102.0	100	105.6	104	101.6	100	96.6	95	101.9
21. FLEUK	97.4	100	102.9	106	95.4	98	93.6	96	97.4
22. FALL	92.5	99	94.6	101	93.0	100	94.1	101	93.4
Regional mean/ Grand mean ≥ 1.00		5		11		12		15	

[a]See Table 5–3 for type of cancer code.
[b]All ratios are multiplied by 100.

central SMSA's. Male leukemia rates increased in the Northcentral, but decreased in the Northeastern SMSA's. Female stomach cancer mortality rates decreased far more in the Northcentral than in the southern metropolises. Female cervical cancer mortality rates decreased less rapidly in the Northeastern than they did in the Southern metropolitan areas. The last and probably the most powerful discriminating disease was female lung cancer. It strongly contrasts the Western with the Northeastern and Northcentral metropolises.

The conclusions derived from the means are supported by the discriminant analysis (Table 5–5). The 14 diseases with the lowest Wilks' lambda values (the lower the value, the stronger the discriminating power; see Appendix 2) include all ten types noted in the initial analysis and male prostate, male all types, female breast, and female ovary. Female stomach and female lung are clearly the most discriminating types of cancer.

In a brief summary, three multiple disease functions were created by a mathematical process (Appendix 2). The first function strongly discriminates between the Southern and Northcentral regions. The second focuses on differences between the Northcentral and the Northeastern regions, and the third on differences between the West and the Northeast.

Table 5–5. Types of Cancer with Strongest Power to Discriminate between Changes in the Cancer Profiles of the 25 Metropolitan Areas in Four Regions

Type of Cancer[a]	Wilks' Lambda	F
1. FSTOM	.52	6.40
2. FLUNG	.56	5.58
3. MLUNG	.66	3.61
4. MALL	.68	3.35
5. MPROS	.71	2.86
6. MLEUK	.72	2.79
7. MPANC	.75	2.39
8. FBRST	.82	1.56
9. FCERV	.82	1.55
10. FOVAR	.82	1.49
11. FALL	.84	1.37
12. MLIV	.85	1.22
13. MRECT	.87	1.06
14. FLEUK	.87	1.06

[a]See Table 5–3 for types of cancer code.

Table 5–6. Southern, Northcentral, and Western Discriminant Functions

	Correlations between Time Change Indices and Function[a]		
Type of Cancer[b]	Southern Function (1)	North Central Function (2)	Western Function (3)
1. FSTOM	.536		
2. FLEUK	.470		
3. FCERV	−.297		
4. FLIV	−.266		
5. MPANC	−.202		
6. MPROS		.473	
7. MLEUK		.463	
8. FBRST	.233	.381	
9. FLUNG			.785
10. MRECT			.407
11. FINT			−.244
12. FPANC			.237

[a]All correlations exceeding 0.200 are shown.
[b]See Table 5-3 for type of cancer code.

The discriminant analysis correctly classified 24 of the 25 SMSA's, which implies distinct regional patterns of change.

Function 1 is called the *southern function* because the Southern metropolitan areas have the highest group mean (2.37) (Tables 5–6 and 5–7). Among the four regions, the Southern SMSA's had the lowest decrease in stomach cancer, the only increase in female leukemia, the highest decrease in female cervical and female liver cancer, the lowest increase in male pancreatic cancer, and the only increase in female breast cancer mortality rates. As its centroid of −1.15 indicates, the Northcentral metropolitan areas exhibited the opposite changes. The Northcentral SMSA's had the highest decrease in female stomach cancer, the second highest decrease in female leukemia, the smallest decrease in cervical cancer, the highest increase in male pancreatic cancer, and second largest

Table 5–7. Canonical Discriminant Functions Evaluated at Region Means

Region	Southern Function (1)	North Central Function (2)	Western Function (3)
Northeast	−0.57	−1.38	−1.06
South	2.37	−0.99	0.33
Northcentral	−1.15	1.44	−0.49
West	−0.11	0.31	2.20

decrease in female breast cancer mortality rates. The Northeastern SMSA's exhibited changes similar to the Northcentral SMSA's, though not as strong. The four Western SMSA's are not strongly discriminated by the time change indices of the six types of cancer in function 1.

Function 2 is called the *northcentral function,* but it just as easily could have been called the *northeast function.* It focuses on three types of cancer in which the rates of change differed in the Northcentral and Northeastern SMSA's. The Northcentral SMSA's had the smallest decrease of male prostate cancer, the Northeast SMSA's the largest decrease. The Northcentral SMSA's had the highest increase in male leukemia rates, and the northeast was the only region to decrease. Finally, the Northcentral metropolises had a slightly larger decrease of female breast cancer mortality rates than the Northeastern metropolises. The six Southern SMSA's had rates of change that were similar to those of the Northeast. Again, the four Western SMSA's are not sharply distinguished by the three diseases of function 2.

Function 3, the *western function,* distinguishes the *four* Western metropolises from the other SMSA's, especially from those in the Northeast. The Western SMSA's had by far the highest increases in female lung, the smallest decrease in male rectal, a modest decrease in female large intestine, and the most pronounced increase in female pancreatic cancer mortality rates. Changes in the Northeastern states were in the opposite direction.

The discriminating power of the time change indices of the 12 diseases is summarized by the observation that 24 of the 25 SMSA's were classified in the actual region by the discriminant functions. The discriminant function scores for the 25 SMSA's allow one to review the identification of each SMSA with the three functions more carefully (Table 5-8). For example, the six Northeastern SMSA's have probabilities ranging from 70 to greater than 99 percent of belonging to the Northeastern group. The Northeastern metropolitan areas identified inversely with the time change indices of all three functions, especially functions 2 and 3. Boston and Paterson, Passaic, and Clifton exhibit strong negative scores for each function. Buffalo and Newark deviate slightly from the Northeastern model. Buffalo exhibits some of the time change indices characteristic of the Southern and Western SMSA's as indicated by its positive scores on functions 1 and 3. However, its strong negative identification with the Northcentral function (function 2) clearly places it in the Northeast group. Newark strongly disassociates with the Southern and Western

Table 5–8. Discriminant Scores of 25 SMSA's for Three Discriminant Functions

Region[a] SMSA	If Incorrect, Projected Group	Probability of Highest Group	Discriminant Scores		
			1—South	2—Northcentral	3—West
NE					
1. BOST		99+	− .45	−2.01	−1.43
2. BUFF		.75	.33	−2.36	.30
3. NY		.81	− .84	− .30	−1.73
4. NWK		.70	−1.89	− .40	−1.23
5. PPC		99+	− .86	−1.75	−1.25
6. PHIL		94	.27	−1.46	−1.01
SO					
7. ATL		99+	4.16	−1.04	1.51
8. BALT	(NE)	81	.04	−1.20	.02
9. DALL		99+	2.86	− .13	−1.24
10. HOUS		99+	3.38	− .65	.96
11. MIAM		95	1.92	−2.05	− .23
12. DC		95	1.84	− .89	.95
NC					
13. CHIC		80	− .77	.72	−1.65
14. CINC		70	−1.56	.48	.64
15. CLEV		99	−1.35	1.60	−1.53
16. DETR		91	− .04	1.14	−1.24
17. KC		99+	−1.88	2.23	−1.12
18. MILW		90	− .31	1.17	−1.95
19. MINN		66	−2.82	.53	1.29
20. PITT		91	− .11	3.60	.92
21. STLO		96	−1.50	1.53	.20
WE					
22. LA		66	.23	−1.34	1.42
23. SD		99+	− .10	.16	2.86
24. SF		87	− .63	2.48	2.22
25. SEAT		99	.06	− .04	2.31

[a]See Table 5–2 for region/city codes.

functions, but less so from the Northcentral function, and therefore there is a small probability that it should be placed in the Northcentral group.

With the exception of Baltimore, all the Southern SMSA's strongly identify with function 1, the Southern function. All but Baltimore have probabilities of at least 95 percent of being in the South. Baltimore, on the other hand, has an 81 percent probability of being in the Northeast, since its time change index profile resembles that of Buffalo and Philadelphia. Baltimore, however, is considered by regional analysts to be on the southern tip of the American manufacturing belt (White, Foscue, and McKnight, 1964), and it possesses the characteristics of a manufacturing city; in contrast to the census bureaus classification of Maryland in the South, the "misclassification" of Baltimore seems quite appropriate.

All nine Northcentral SMSA's were correctly classified. Six of the nine had probabilities of 90 percent or greater of belonging to the Northcentral group. Chicago, Cincinnati, and Minneapolis/St. Paul are the three with probabilities ranging from 66 to 80. Cincinnati and Minneapolis/St. Paul, as expected, identify negatively with the southern function and positively with the northcentral function. But both have small probabilities of belonging to the western function because of their positive identification with function 3.

All four Western SMSA's strongly identify with function 3, the western function. However, Los Angeles and San Francisco also identify with another function. Los Angeles has a strong negative association with the northcentral function and San Francisco a strong positive association with the northcentral function. Accordingly, both have slight probabilities of belonging to other groups.

Although their rates of change seem similar, the time change indices of 12 types of cancer strongly differentiate between the most populous Northeastern, Northcentral, Southern and Western SMSA's. The Southern metropolises experienced the smallest decreases in stomach cancer, the only increase in female leukemia, the sharpest decrease in female cervical and female liver cancer, the smallest increase in male pancreatic cancer, and the only increase in female breast cancer mortality rates during the period 1950–69. Atlanta, Dallas, and Houston exemplify the changing Southern metropolitan cancer profile.

The Northcentral SMSA's are distinguished by the relatively small decrease in white male prostate cancer and the high increase in white male leukemia mortality rates. Cleveland and Kansas City exemplify the time change indices of the Northcentral states.

The four Western SMSA's, exemplified by San Diego and Seattle, experienced extremely large increases in female lung and female pancreatic cancer mortality rates and the smallest decrease in male rectal cancer mortality rates.

The six Northeastern SMSA's identify negatively with the time change profiles of the other three regions. Boston and Paterson, Passaic, and Clifton epitomize metropolitan regions that had the most modest increases of every type of major cancer during the two decades.

CHILD AND TEENAGE CANCER MORTALITY RATES
IN URBAN AMERICA

The analyses presented in Chapters 4 and 5 demonstrate that urban areas have been characterized by higher total population cancer mortality rates than rural areas. Is this conclusion also true for the youngest populations? There are some reasons to suspect that there may not be an urban/rural difference in child and teenage cancer. First, the major types of cancer afflicting children and teenagers (leukemia, brain and other central nervous system cancers, and lymphoma) exhibited relatively small urban/rural differences for the entire population in comparison to respiratory, digestive, and urinary cancers. Second, child and teenage cancer mortality rates have been decreasing, which suggests different etiologies as well as improved survival rates for the young. Third, initial epidemiological studies suggest that rural as well as urban occupations are associated with higher child and teenage cancer incidence (Hemminki et al., 1981).

Before child and teenage cancer mortality rates in highly urbanized and less urbanized areas of the United States can be compared, it is necessary to briefly consider some questions of data. A major conclusion to be drawn from scanning leukemia and all other cancer mortality data for the 0- to 19-year-old populations is that there has been a dramatic decline in rates, especially during 1970–75 and among the white population (Figures 5–3 through 5–5). Although rates began to decline sharply during the late 1960's, the decline in child and teenage cancer mortality rates between 1970–75 is so marked that it was decided to analyze the 1950–69 data in this chapter. Had aggregate rates for 1950–75 been calculated it was feared that important differences between counties might be lost.

Four steps were taken to secure cancer rates with the lowest possible standard deviations. Age-adjusted rates were calculated for the entire period 1950–69 and for the entire population 0–19, rather than for each 5- or 10-year time period and for segments of the population. The analyses were made only for the white population because few counties have enough nonwhite cancer deaths to produce meaningful results. Finally, the almost 300 counties that were at least 75 percent urban were used as the urban counties rather than the 25 SMSA's in order to increase the population base. Another methodological issue was what to use as the base line area when comparing the urban counties and the entire United States. The advantages of using the United States as the base line are consistency; whether the base line is the nation or the remainder of the

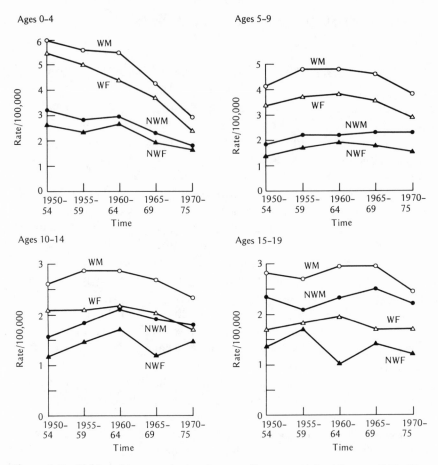

Figure 5–3. Child and teenage leukemia mortality rates, United States, 1950–75

nation has little impact upon the results that follow, with one exception. The exception is the comparison between the aggregate of the 293 urban counties and the United States because more than one-half of the American population resides in the 293 counties. Both the nation and the remainder of the nation are used as base lines for this comparison (Table 5–9).

Comparisons between the most urbanized counties and the entire United States show that the strongly urban counties as a whole have higher rates of child and teenage cancer mortality rates than the entire nation (Table 5–9). The most pronounced difference is 7 percent for male and 9 percent for female brain and other central nervous system cancer.

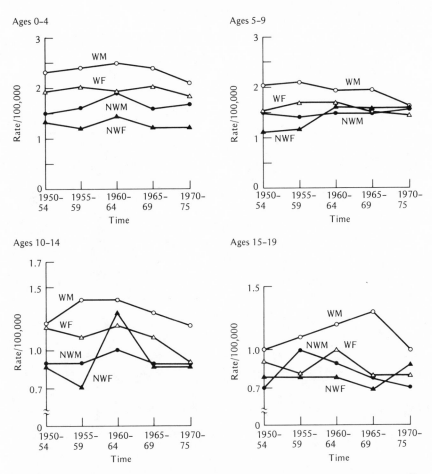

Figure 5–4. Child and teenage brain and central nervous system cancer mortality rates, United States, 1950–75

Brain and other central nervous system cancer, as well as male and female leukemia, male and female all types, male non-Hodgkin's lymphoma, and male bone cancer have rates of child and teenage cancer mortality rates in the strongly urban counties that are significantly higher than the national rates at the 95 percent confidence level.

Of the 16 rates presented, only male kidney cancer has a lower rate in the most urbanized counties than in the nation. It is particularly interesting to note that the strongly urban/U.S.A. ratios are sometimes higher for 0- to 19-year-old populations than for the total population.

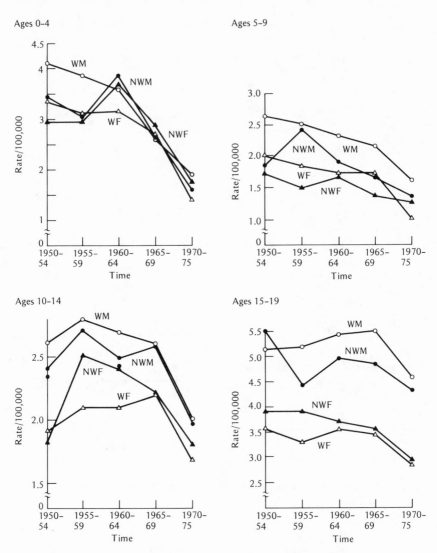

Figure 5–5. Child and teenage all other types of cancer mortality rates, United States, 1950–75

Having demonstrated that child and teenage cancer mortality rates are usually higher in the most urbanized counties of the United States than they are in the entire nation, a search was made for regional patterns of child and teenage cancer mortality. Preliminary analyses revealed a data problem. The rates of the less populated counties have very large stan-

Table 5–9. Comparison of Child and Teenage Cancer Mortality Rates in 293 Strongly Urban and all Counties in the Continental United States, 1950–69

		White Male					White Female			
Type of Cancer	Age-adj. 0–19 rate per 100,000	0–19 SU rate[a] U.S.A. rate	Sign at .05?[b]	0–19 SU rate REMUS[c]	Age-adj. total pop. SU rate U.S.A. rate	Age-adj. 0–19 rate per 100,000	0–19 SU rate U.S.A. rate	Sign. at .05?	0–19 SU rate REMUS[c]	Age-adj. total pop. SU rate U.S.A. rate
1. All types	9.59	104	Yes	110	109	7.55	105	Yes	111	105
2. Leukemia	4.20	104	Yes	110	101	3.36	105	Yes	111	102
3. Brain/central nervous system	1.90	107	Yes	117	106	1.58	109	Yes	122	107
4. Non-Hodgkin's lymphoma	0.92	105	Yes	110	107	0.43	103	No	106	108
5. Bone	0.53	105	Yes	111	100	0.41	101	No	104	95
6. Kidney	0.38	99	No	98	109	0.43	102	No	104	103
7. Hodgkin's disease	0.33	104	No	111	103	0.19	105	No	112	106
8. All other	1.33	102	No	104	110	1.15	100	No	100	105

[a]SU, strongly urban counties; all ratios have been multiplied by 100.

[b]If the strong urban set has a 95 percent or higher probability of having a higher rate than the nation, then a yes is indicated.

[c]REMUS, remainder of the United States.

159

dard errors, a result expected from a combination of low rate cancer and small populations at risk. This data problem was solved by working with increasingly more populous counties until the results were not seriously changed. This point was reached when the county populations exceeded 400,000 whites. The final set of results are derived from 81 counties with white populations exceeding 400,000 in 1960. These 81 counties contained 41 percent of the white population of the United States in 1970.

The 81 counties are spread across the United States with the largest concentrations in the Northeast and in California (Figure 5–6). The dispersed geographical distribution limits generalizations about regional patterns of child and teenage disease. Nevertheless, statistical analyses were made to spot obvious geographical patterns.

The counties with the highest total cancer mortality rates among those 0–19 tend to be in New York/New Jersey and northern California. The five-county New York City aggregate and neighboring, Nassau, New York; Essex, Hudson, and Union counties, New Jersey had male rates between 13 and 19 percent higher than the national average (Table 5–10). San Francisco and adjacent Contra Costa County, California also had rates significantly higher than the national average for white males 0 to 19 years old during 1950–69. Two of the lowest total male rates were observed in Massachusetts and two in Florida.

Figure 5–6. Location of the 81 counties for child and teenage cancer mortality study

Table 5–10. Highest and Lowest Total Cancer Mortality Rates Among the 81 Most Urbanized and the Most Populous Counties of the United States, Age Group 0–19, Time Period 1950–69

White Male, Highest Rate Counties	Rate/100,000	County Rate[a] U.S.A. Rate	Signif.[b]
Suffolk, Mass.	12.7	137	99
Multnomah, Ore.	11.5	125	99
San Francisco, Calif.	11.3	123	95
New York City, N.Y.[c]	11.0	119	99
Essex, N.J.	10.9	118	99
Jackson, Mo.	10.7	116	90
Nassau, N.Y.	10.7	116	99
Contra Costa, Calif.	10.6	115	90
Hudson, N.J.	10.5	115	90
Lucas, Ohio	10.5	114	80
Union, N.J.	10.5	114	80
Philadelphia, Pa.	10.4	113	95
White Female, Highest Rate Counties			
San Francisco, Calif.	10.1	140	99
Suffolk, Mass.	9.8	136	99
Montgomery, Md.	9.6	133	99
Contra Costa, Calif.	9.1	126	99
Bergen, N.J.	9.0	124	99
Baltimore City, Md.	8.9	123	95
Philadelphia, Pa.	8.8	122	99
Multnomah, Ore.	8.8	122	95
Sacramento, Calif.	8.8	121	95
Dade, Fla.	8.8	121	95
Montgomery, Ohio	8.6	120	95
DuPage, Ill.	8.6	119	90

White Male, Lowest Rate Counties	Rate/100,000	County Rate U.S.A rate	Signif.[b]
Broward, Fla.	6.1	67	99
Baltimore County, Md.	7.5	81	99
Bristol, Mass.	8.1	88	80
Essex, Mass.	8.4	91	80
Riverside, Calif.	8.4	91	—
Jefferson, Ky.	8.6	93	—
Maricopa, Ariz	8.6	93	80
Camden, N.J.	8.6	93	—
Lake, Ind.	8.6	93	—
Fairfield, Conn.	8.6	94	—
Allegheny, Pa.	8.6	94	—
Duval, Fla.	8.7	94	—
White Female, Lowest Rate Counties			
Hillsborough, Fla.	5.5	76	95
Riverside, Calif.	5.7	79	95
Bristol, Mass.	5.9	82	95
Jackson, Mo.	6.0	83	95
Essex, Mass.	6.1	85	95
Duval, Fla.	6.2	86	90
Delaware, Pa.	6.2	86	90
Hamilton, Ohio	6.2	86	95
Jefferson, Ala.	6.4	88	80
Maricopa, Ariz.	6.4	88	90
Hudson, N.J.	6.4	89	80
Baltimore County, Md.	6.5	90	80

[a] All ratios have been multiplied by 100.

[b] Probability that the county rate is higher or lower than national rate if it is at least 80.

[c] All five New York City counties were merged by the National Cancer Institute.

With one exception, the highest female rates did not exhibit a distinctive pattern. The single exception is northern California. Three northern California counties had rates significantly higher than the entire nation. And again, two Florida and two Massachusetts counties had among the lowest female rates.

More detailed studies using multivariate analyses and mapping revealed multiple disease, high rate areas in the New York City/northern New Jersey, New England, northern California, and Great Lakes areas. Beginning with the New York City/northern New Jersey area, eight counties and the New York City aggregate had child and teenage cancer mortality leukemia rates that averaged 13 percent higher than the national total (Table 5–11). Six of the counties and New York City had male rates that had an 80 percent or greater probability of being higher than the national rate.

The differences between the New York/northern New Jersey area and the United States are even more marked for Hodgkin's disease. During 1950–69, the average male and female Hodgkin's disease cancer mortality rate in these counties was almost 40 percent higher than the national rate for those 0 to 19 years old. Further research led to the observations that other counties in the New York City/New Jersey/Philadelphia corridor also had relatively high rates of leukemia and Hodgkin's disease and that non-Hodgkin's lymphoma rates were also elevated in this area (Greenberg et al., 1980a).

Like the New York/New Jersey counties to the south, ten New England counties had relatively high rates of Hodgkin's disease during 1950–69 (Table 5–12). Specifically, seven of the ten had white male rates higher than the nation, and eight of the ten had female rates higher than the nation. However, due to the limited number of deaths from this disease, only one, Suffolk County (Boston), had a rate statistically significantly higher than the nation as a whole. These same ten counties also exhibited relatively high rates of child and teenage female, but not male bone cancer mortality.

Six heavily populated and urbanized northern California counties manifested clearly higher female leukemia and to a lesser extent, male leukemia rates than the entire nation (Table 5–13). In contrast, the five southern California counties did not exhibit unusually high rates of leukemia among children and teenagers during 1950–69.

The last interesting regional agglomeration of relatively high rates was found in the Great Lakes region. In that region, 12 of the 15 counties had male, non-Hodgkin's lymphoma cancer mortality rates exceeding the

Table 5–11. Selected Child and Teenage Cancer Mortality Rates, New York/New Jersey's Most Urbanized and Populous Counties, 1950–69

County	White Male Leukemia		White Female Leukemia		White Male Hodgkin's		White Female Hodgkin's	
	C rate[a] U.S.A. rate	Signif.[b]	C rate U.S.A. rate	Signif.[b]	C rate U.S.A. rate	Signif.[b]	C rate U.S.A. rate	Signif.[b]
Bergen, N.J.	114	80	143	99	111	—	107	—
Essex, N.J.	125	99	118	80	181	90	222	90
Hudson, N.J.	106	—	91	—	152	—	147	—
Passaic, N.J.	104	95	108	—	103	—	91	—
Union, N.J.	121	99	105	—	144	—	106	—
Nassau, N.Y.	124	99	101	—	153	90	88	—
New York City, N.Y.[c]	119	99	116	99	158	99	149	80
Suffolk, N.Y.	113	95	113	90	158	—	213	80
Westchester, N.Y.	110	80	111	—	51	—	166	—

[a]C, county cancer mortality rate; all ratios have been multiplied by 100.
[b]Probability of county rate being higher than the national rate if the probability is at least 80.
[c]All five New York City counties were merged by the National Cancer Institute.

Table 5–12. Child and Teenage Hodgkin's Disease and Female Bone Cancer Mortality Rates in New England's Most Urbanized and Populous Counties, 1950–69

County	White Male Hodgkin's		White Female Hodgkin's		White Female Bone	
	C rate[a] / U.S.A. rate	Signif.[b]	C rate[a] / U.S.A. rate	Signif.[b]	C rate[a] / U.S.A. rate	Signif.[b]
Fairfield, Conn.	130	—	100	—	104	—
Hartford, Conn.	125	—	242	—	44	—
New Haven, Conn.	98	—	102	—	135	—
Bristol, Mass.	66	—	191	—	68	—
Essex, Mass.	142	—	52	—	100	—
Hamaden, Mass.	132	—	146	—	126	—
Middlesex, Mass.	71	—	123	—	121	—
Norfolk, Mass.	135	—	55	—	135	—
Suffolk, Mass.	213	95	135	—	134	—
Providence, R.I.	111	—	166	—	197	90

[a]C, county cancer mortality rate; all ratios have been multiplied by 100.
[b]Probability of county rate being higher than the national rate if the probability is at least 80.

Table 5-13. Child and Teenage Leukemia Cancer Mortality Rates in Northern California's Most Urbanized and Populous Counties, 1950-69

County	White Male C rate[a] / U.S.A. rate	Signif.[b]	White Female C rate[a] / U.S.A. rate	Signif.[b]
Alameda	101	—	123	95
Contra Costa	117	80	127	95
Sacramento	113	80	122	90
San Francisco	127	95	150	99
San Mateo	96	—	139	99
Santa Clara	111	80	113	—

[a]C, county cancer mortality rate; all ratios have been multiplied by 100.
[b]Probability of county rate being higher than the national rate if the probability is at least 80.

Table 5-14. Child and Teenage Non-Hodgkin's Lymphoma Cancer Mortality Rates in the Great Lakes' Most Urbanized and Populous Counties, 1950-69

County	White Male C rate[a] / U.S.A. rate	Signif.[b]	White Female C rate[a] / U.S.A. rate	Signif.[b]
Cook, Ill.	126	95	101	—
DuPage, Ill.	114	—	119	—
Lake, Ind.	80	—	111	—
Marion, Ind.	103	—	116	—
Jefferson, Kty.	100	—	176	80
Macomb, Mich.	116	—	118	—
Oakland, Mich.	120	—	171	90
Wayne, Mich.	90	—	65	—
Cuyahoga, Ohio	128	90	82	—
Franklin, Ohio	115	—	151	—
Hamilton, Ohio	112	—	97	—
Lucas, Ohio	94	—	69	—
Montgomery, Ohio	123	—	221	95
Summit, Ohio	170	95	134	—
Milwaukee, Wisc.	118	—	92	—

[a]C, county cancer mortality rate; all ratios have been multiplied by 100.
[b]Probability of county rate being higher than the national rate if the probability is at least 80.

nation as a whole (Table 5–14). Three counties had male rates statistically significantly higher than the entire nation at the .10 or more level. Ten of the 15 counties had female non-Hodgkin's lymphoma rates higher than the nation's. Briefly recapitulating, during 1950–69, portions of the northeast, northern California, and the Great Lakes region each exhibited relatively high mortality rates of specific types of child and teenage cancer.

A regional perspective may not be the best means of finding clues about the spatial distribution of child and teenage cancer. Perhaps, socioeconomic status, ethnicity, economic profile, and other characteristics are better clues to the patterns than regional location. Accordingly, 13 indicators of socioeconomic status, occupation, ethnicity, and population migration for the year 1960 were gathered for the 81 counties and correlated with cancer rates for 1950–69 for those aged 0–19. The correlations were weak, few were significant at the .01 confidence level, and those that were significant became insignificant when the correlations were repeated with a sample drawn from the 81 counties.

Overall, the geographical distribution of child and teenage cancer mortality during 1950–69 in urban areas shows regional differences. However, there are no strong clues to explain the child and teenage pattern. Ecological and epidemiological studies may be helpful in finding clues (e.g., Greenberg, 1980b).

SUMMARY

This chapter began with three questions, the answers to which will be briefly summarized:

1. Do the metropolitan areas of the Northeast, Northcentral, South, and West differ in cancer mortality profiles?

As expected, during 1950–54 the urbanized and industrialized Northeastern region had the highest rates of cancer mortality, followed by the Northcentral region. The Southern and Western regions had lower rates. The most marked differences were for digestive tract cancers: rectum, large intestine, esophagus, and female stomach. By 1970–75, with a few exceptions, these differences, many of which had exceeded 15 percent, narrowed to less than 7 percent. The most interesting example was white female respiratory cancer. Female rates of the respiratory diseases increased far more rapidly in the Southern and Western urban areas than in the Northeast and Northcentral urban areas.

2. Are there differences between the metropolises of the United States and the entire nation in child and teenage cancer mortality rates?

The most urbanized counties were found to have significantly higher child and teenage (0–19) cancer rates for all types and brain and central nervous sytem cancers and leukemia, male non-Hodgkin's lymphoma, and male bone cancer during 1950–69. It is noteworthy that the strongly urban/U.S.A. ratios were often higher for the child and teenage population than for the total population as measured by age-adjusted rates for the 0- to 19-year-old population.

3. Within the group of most urbanized areas, are there regional differences in child and teenage cancer mortality profiles?

Among the approximately 300 most urbanized counties, counties with the highest rates for those 0 to 19 years old tend to be in the New York/New Jersey region and northern California. The New York City/Northern New Jersey area had relatively high child and teenage leukemia, Hodgkin's disease, and to a lesser extent, non-Hodgkin's lymphoma rates. Ten strongly urban counties in New England (in Massachusetts, Connecticut, and Rhode Island) exhibited relatively high rates of Hodgkin's disease. Six northern California urban counties manifested high child and teenage leukemia rates. The last interesting regional pattern was the relatively high rates of non-Hodgkin's lymphoma in the urbanized counties near the Great Lakes (in Illinois, Indiana, Kentucky, Michigan, Ohio, and Wisconsin).

REFERENCES

Greenberg, M., Caruana, J., Holcomb, B., Greenberg, G., Parker, R., Louis, J., and White, P. 1980a. "High cancer mortality rates from childhood leukemia and young adult Hodgkin's disease and lymphoma in the New Jersey–New York–Philadelphia Metropolitan Corridor." *Can. Res.* 40: 439–443.

Greenberg, M., Preuss, P., and Anderson, R. 1980b. "Clues for case control studies of cancer in the Northeast Urban Corridor." *Soc. Sci. Med.* 14D: 37–43.

Hemminki, K., Saloniemi, I., Salonen, T., Partanen, T., and Vainio, H. 1981. "Childhood cancer and parental occupation in Finland." *J. Epidemiol. Comm. Health* 35: 11–15.

White, C., Foscue, E., and McKnight, T. 1964. *Regional Geography of Anglo-America,* Englewood Cliffs: Prentice-Hall.

6 Cities and Suburbs: Toward a Homogeneous Profile

The literature on cities and suburbs is too voluminous (see, for example, Masotti and Hadden, 1973; Berry, 1973) to review here, but some particularly salient points will be highlighted. The urban/industrial city developed because an increasingly interdependent society needed nodes of high interaction potential. These engendered high densities, high rents, and the animosity of many who preferred an agrarian American (see, for example, Wirth, 1938).

Suburbanization is of long-standing importance in the urbanization process in America. Since the early 1950's, it has been accelerated by modern transportation and communication systems, which have reduced the need for concentration. As early as 1900, the suburbs were viewed as a panacea for the ills of the city (Wells, 1902). H. G. Wells predicted that the large urban regions of the 1960's and 1970's would reach out over 100 miles, and hoped that they would be more like "town provinces" with space for privacy and for natural vistas.

Whereas Wells had hoped for a different and better quality of life in the suburbs, what has emerged is a city-suburb. As Oscar Handlin (1966), the historian bluntly put it: "The differences between city and country have been attenuated almost to the vanishing point." Employment opportunities and population densities of central cities have decreased, as almost all growth in the United States has been in suburbs (Brunn and Wheeler, 1980). Adult and juvenile crime, shortages of recreational space, and many other perceived ills of cities are increasingly characteristic of suburbs.

This chapter compares cancer mortality rates in the largest central cities and their suburbs.

Three questions are posed:

1. How do cancer mortality rates in central cities and suburbs compare with those of less urbanized areas of the United States?
2. How do central city and suburban cancer mortality profiles differ?
3. Do the cancer mortality profiles of amenity retirement areas reflect the rural areas in which they are situated or the cities and suburbs from which much of their population came?

DATA BASE AND METHODS

The potential data base is large, but it imposes two important limitations on the researcher: geographical detail and the reliability of cancer mortality rates. Ideally, geographical detail should be fine enough to allow the use of U.S. Census Bureau definitions of central cities and suburbs of Standard Metropolitan Statistical Areas (SMSA's). Practically, however, SMSA's had to be disaggregated into approximate central city and suburbs following county boundaries. The counties containing the central cities were called the central city counties; all other counties were called suburbs. As noted below, this limited the comparisons to populous, multiple county SMSA's.

The inability to reproduce exactly the Census Bureau's central city and suburban classification exacerbates the problem of reliability of mortality rates. Together, these limitations led to the following decisions:

1. Twenty-two of the 25 most populated SMSA's in 1970 were selected for the central city and suburban comparisons for 1950–75 (Table 6–1). These were chosen because county geographical detail and population size were inadequate for the smaller SMSA's. Three of the 25 most populated SMSA's were excluded because their central cities and suburbs are in one county (Los Angeles, Miami, and San Diego). Despite this conservative step, the comparison between central cities and suburbs is somewhat blurred by the use of county data.
2. Comparisons were limited to white male and white female populations because there were not enough nonwhites, especially in the suburbs.
3. Although records are available for 32 types of cancer in white males and 33 types in white females, comparisons between individual cit-

Table 6–1. Standard Metropolitan Statistical Areas Included in the Central City and Suburb Analysis

| | Number of Counties | |
SMSA	Central City	Suburb
1. Atlanta, Ga.	2	3
2. Baltimore, Md.	1	5
3. Boston, Mass.	1	4
4. Buffalo, N.Y.	1	1
5. Chicago. Ill.	2	4
6. Cincinnati, Ohio	1	6
7. Cleveland, Ohio	1	3
8. Dallas, Tex.	1	5
9. Detroit, Mich.	1	2
10. Houston, Tex.	3	2
11. Kansas City, Kans.–Mo.	1	5
12. Milwaukee, Wisc.	2	2
13. Minneapolis, St. Paul, Minn.	2	3
14. New York, N.Y.	5	4
15. Newark, N.J.	1	2
16. Paterson/Passaic/Clifton, N.J.	1	1
17. Philadelphia, Pa.	1	7
18. Pittsburgh, Pa.	1	3
19. St. Louis, Mo.	1	6
20. San Francisco, Calif.	1	4
21. Seattle, Wash.	1	1
22. Washington, D.C.	1	6
Total	32	79

ies and suburbs were limited to major groups (e.g., digestive, respiratory) or to total cancer mortality in order to increase the reliability of the comparisons.

4. Some comparisons were made between five-year time periods; others were made between ten-year time periods (e.g., 1950–59), again to increase the reliability of the comparisons.

The final data base from which the mortality rates were calculated for 1950–75 consists of 6,525,000 deaths in the white population: 1,385,000 or 21 percent in the 25 central city counties (32 counties); 754,000 or 12 percent in the suburbs of these central city counties (79 counties); and 4,386,000 or 67 percent in the remainder of the continental United States (about 2,900 counties). Although this set of data is imperfect, it is based on enough records and on a long enough time span to determine the extent to which cancer mortality patterns have changed within metropolitan regions.

AN INITIAL PERSPECTIVE: CENTRAL CITY
AND SUBURBAN AGGREGATES

The results strongly support the view that central city rates were formerly much higher than suburban rates, but that the gap between the two has narrowed substantially. During 1950–54, the total white male cancer mortality rate in the aggregated central city counties was 17 percent higher than in the suburbs (Figure 6–1; Table A1–47). By 1960–64 the difference had dropped to 12 percent, and by 1970–75, to 5 percent. For white females, the difference dropped from 11 to 3 percent during the two decades. This drop has been gradual (Table A1–49; Figure 6–2).

Although there are many cities in the remainder of the nation, there are also over 2,000 counties in which the rural population exceeds one-

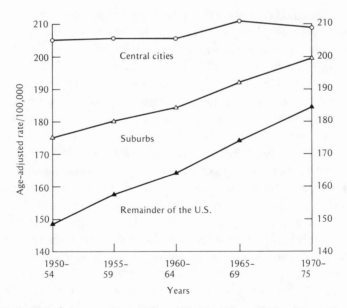

Figure 6–1. Trends in cancer mortality, central cities, suburbs, remainder of the United States, white males, 1950–75

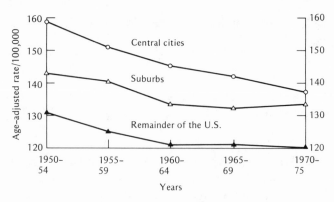

Figure 6–2. Trends in cancer mortality, central cities, suburbs, remainder of the United States, white females, 1950–75

half of the county population. As expected, therefore, both the central city and the suburban counties have had much higher cancer mortality rates than the remainder of the nation. These differences are clearly decreasing, especially among white males, as illustrated by Figures 6–1 and 6–2.

A more detailed review confirms the expectations about mortality rates for cancer at specific body sites as set forth throughout this volume. The biggest gaps between the central cities and the suburbs during 1950–54 were in rates for digestive and respiratory system cancer (Tables A1–47 through A1–50). Central city/suburb ratios for the following seven cancer sites were above 1.20: male and female lung; male and female stomach; male and female rectum; male tongue/mouth; male esophagus; male larynx; and female kidney. Also as expected, the rate for nearly every other cancer was also higher in the central city counties during the early 1950's. Indeed, only in the cases of cancer of the prostate, male non-melanoma skin, and corpus uteri and of female melanoma was the aggregate central city rate less than the aggregate suburban rate. Thus, during the early 1950's central cities as a whole had much higher respiratory and digestive system and total cancer mortality rates than their suburbs.

By 1970–75 the picture had changed. Although aggregate central city rates continued to be higher, differences between the central cities and the suburbs substantially decreased (Figures 6–1 and 6–2). The most interesting observation is the dramatic narrowing of the gap in both male and female respiratory cancer (Tables A1–47 and A1–49). This group of diseases increased substantially as a cause of death in both the central cities and the suburbs, as well as in the United States as a whole. In the major metropolitan regions, the most pronounced increases were in the suburbs. Both male and female rates almost doubled during the two decades. Digestive system cancer mortality rates decreased, though more so in the central cities than in the suburbs, leading to a narrowing of the gap between the central city and suburban aggregates.

A FINER PERSPECTIVE: TWENTY-ONE PAIRS
OF CENTRAL CITIES AND SUBURBS

The second step was to compare the central city and suburban rates of each of the SMSA's. This comparison was necessary because the five most populous SMSA's among the twenty-two (New York, Chicago, Philadelphia, Detroit, and San Francisco) contain almost one-half of the population at risk of the 22 SMSA's. Accordingly, the aggregate central city and suburbs compared above might be overly influenced by these five. The Houston SMSA had to be eliminated from this comparison because 93 percent of its population resided in the central city counties in comparison to an average of 59 percent in the other 21 SMSA's. The suburban population base of Houston was judged too small; therefore the results that would have been obtained if it had been included might have been misleading. In response to the rate reliability limitation discussed above, the initial comparisons were limited to total white male and female cancer mortality and to 1950–59, 1960–69, and 1970–75.

The results support the conclusions drawn from the aggregate analysis. With respect to central cities and the remainder of the United States, all 21 central cities had a higher white male rate than the remainder of the United States during 1950–59 (Table 6–2). The median central city rate was 32 percent higher with a range of 48 percent higher (Boston) to 10 percent higher (Dallas). Nineteen of 21 central cities had a white female total cancer mortality rate higher than the remainder of the nation (Table 6–3). The median rate was 18 percent higher. Atlanta and Dallas were the exceptions insofar as they had lower white female rates than the remainder of the nation during 1950–59. New York City exhibited the highest rate, 31 percent higher than the remainder of the nation.

The suburbs of the 21 central cities almost always had higher rates than the rest of the United States. Specifically, 17 of the 21 suburbs had higher total white male cancer mortality rates with a range of 32 percent higher (Paterson/Passaic/Clifton) to 19 percent lower (Minneapolis/St.

Table 6-2. White Male, All Types Cancer Mortality Rates, 1950–75, 21 Central Cities and Suburbs

	Central Cities						Suburbs					
	1950–59		1960–69		1970–75		1950–59		1960–69		1970–75	
SMSA	Rate	Rank	Rate	Rank	Rate	Rank	Rate	Rank	Rate	Rank	Rate	Rank
1. Atlanta	171.1	20	183.6	19	190.6	20	132.4	20	162.4	17	189.5	16
2. Baltimore	222.0	2	247.7	1	257.5	1	174.6	7	183.4	13	210.9	3
3. Boston	226.1	1	221.7	4	233.5	3	185.5	4	192.4	5	205.6	6
4. Buffalo	201.9	13	211.1	11	216.5	8	181.4	5	185.8	10	212.1	2
5. Chicago	202.3	11	205.8	14	203.9	15	168.4	12	174.4	14	189.1	17
6. Cincinnati	192.0	16	214.7	8	225.6	5	172.2	10	199.7	3	222.1	1
7. Cleveland	205.7	9	217.0	6	215.6	9	155.1	16	174.1	15	196.2	10
8. Dallas	168.6	21	185.0	18	204.9	14	138.8	18	157.7	19	172.6	20
9. Detroit	202.0	12	214.9	7	211.0	10	172.7	9	187.7	8	195.8	11
10. Kansas City	173.8	18	186.6	17	210.1	11	132.9	19	155.8	21	181.1	19
11. Milwaukee	201.8	14	199.5	16	196.1	18	158.5	15	160.0	18	182.0	18
12. Minneapolis/St. Paul	172.6	19	181.7	21	188.4	21	128.2	21	155.8	20	164.7	21
13. New York	218.0	3	212.8	9	206.2	13	190.8	3	200.0	2	204.3	7
14. Newark	217.3	4	211.8	10	202.0	16	196.7	2	193.6	4	208.8	4
15. Paterson/Passaic/Clifton	211.2	7	207.8	13	210.0	12	201.2	1	202.7	1	206.2	5
16. Philadelphia	216.8	5	225.9	3	226.5	4	180.5	6	188.7	6	197.0	8
17. Pittsburgh	195.0	15	208.3	12	218.6	7	174.4	8	188.4	7	193.1	12
18. St. Louis	214.9	6	227.0	2	250.4	2	171.6	11	185.1	11	192.5	13
19. San Francisco	204.3	10	219.4	5	221.3	6	167.8	13	183.9	12	189.8	15
20. Seattle	178.9	17	182.6	20	196.3	17	152.2	17	172.1	16	190.6	14
21. Washington, D.C.	206.6	8	201.7	15	195.7	19	165.4	14	187.2	9	196.6	9
Highest	226.1	1	247.7	1	257.5	1	201.2	1	202.7	1	222.1	1
Median	202.3	11	211.1	11	210.1	11	171.6	11	185.1	11	195.8	11
Lowest	168.6	21	181.7	21	188.4	21	128.2	21	155.8	21	164.7	21

Table 6-3. White Female, All Types Cancer Mortality Rates, 1950–75, 21 Central Cities and Suburbs

	Central Cities						Suburbs					
	1950–59		1960–69		1970–75		1950–59		1960–69		1970–75	
SMSA	Rate	Rank	Rate	Rank	Rate	Rank	Rate	Rank	Rate	Rank	Rate	Rank
1. Atlanta	126.5	20	123.4	20	118.2	20	105.4	21	105.1	20	114.0	20
2. Baltimore	158.7	3	155.4	1	147.3	2	134.4	15	126.4	15	131.7	9
3. Boston	157.9	4	144.7	6	143.9	4	146.2	3	132.2	8	132.5	8
4. Buffalo	149.4	13	136.7	15	137.2	11	136.6	12	134.9	6	137.2	5
5. Chicago	153.7	7	143.0	10	137.5	10	136.4	13	128.4	14	125.9	14
6. Cincinnati	153.9	6	143.4	9	146.8	3	143.7	5	140.0	4	141.1	2
7. Cleveland	150.1	12	143.7	8	132.5	14	141.6	8	137.0	5	129.4	11
8. Dallas	124.2	21	117.5	21	117.0	21	108.6	20	104.8	21	104.2	21
9. Detroit	148.7	14	138.0	12	130.7	16	142.5	7	130.3	9	130.7	10
10. Kansas City	132.4	19	128.2	17	130.7	15	130.5	18	111.5	19	117.2	19
11. Milwaukee	151.5	11	136.7	14	126.7	17	140.3	9	124.5	16	121.9	18
12. Minneapolis/St. Paul	140.7	17	125.9	18	124.1	19	123.0	19	121.4	17	124.1	17
13. New York	167.7	1	153.3	2	142.9	6	145.9	4	142.9	1	144.6	1
14. Newark	157.0	5	152.2	3	142.4	7	151.6	2	141.8	3	136.9	6
15. Paterson/Passaic/Clifton	152.6	9	144.5	7	141.5	8	155.8	1	142.6	2	138.8	3
16. Philadelphia	160.6	2	150.7	4	138.9	9	143.3	6	133.8	7	133.1	7
17. Pittsburgh	146.9	16	139.9	11	135.1	13	138.2	10	129.5	12	124.5	15
18. St. Louis	152.7	8	137.8	13	135.2	12	133.0	16	128.7	13	127.7	13
19. San Francisco	151.8	10	147.7	5	152.3	1	135.0	14	130.2	10	137.9	4
20. Seattle	139.5	18	125.8	19	125.3	18	130.5	17	120.5	18	124.2	16
21. Washington, D.C.	147.0	15	136.0	16	143.9	5	137.5	11	129.5	11	128.3	12
Highest	167.7	1	155.4	1	152.3	1	155.8	1	142.9	1	144.6	1
Median	151.5	11	139.9	11	137.2	11	137.5	11	129.5	11	129.4	11
Lowest	124.2	21	117.5	21	117.0	21	105.4	21	104.8	21	104.2	21

Paul). In addition to Minneapolis/St. Paul, three other suburbs had lower total cancer mortality rates than the remainder of the nation: Atlanta, Kansas City, and Dallas. Eighteen of the 21 suburbs had a higher white female cancer mortality rate than the rest of the nation. The median rate was 7 percent higher with a range from 22 percent higher (Paterson/Passaic/Clifton) to 21 percent lower (Atlanta). Minneapolis/St. Paul and Dallas were the other two suburbs with white female rates that were lower than the remainder of the nation.

In summary, during 1950–59, the central cities typically had total cancer mortality rates 20 to 30 percent higher than the remainder of the nation, whereas suburban rates were usually about 10 percent higher. The few central cities and suburbs (Atlanta, Dallas, Kansas City and Minneapolis/St. Paul) with lower total cancer mortality rates than the remainder of the United States were in the South and Midwest.

The same proportion of central cities and suburbs had higher rates than the remainder of the nation two decades later. However, the gap between, on the one hand, the central cities and the suburbs and, on the other, the remainder of the United States closed. As indicated below (Table 6–4), the difference between the median total cancer mortality rates and the remainder of the nation declined, with the exception of white female suburban rates, which remained about the same.

The central cities not only had higher cancer mortality rates than the remainder of the United States, but the central cities also had higher rates of cancer mortality than their suburbs (Table 6–5). Every one of the central cities had a higher white male rate than its suburbs during 1950–59. The median central city rate was 21 percent higher than the median suburban rate. At one end of the spectrum, the central city counties of the

Table 6–4. Comparison of Total Cancer Mortality Rates: Twenty-One Central Cities, Suburbs, and Remainder of the Nation, 1950–59 and 1970–75

| | Median of 21 Central Cities or Suburbs/Remainder of Nation | | | |
| | White Male | | White Female | |
Time Period	Central Cities	Suburbs	Central Cities	Suburbs
1950–59	1.32	1.12	1.18	1.07
1970–75	1.14	1.06	1.14	1.08

Table 6—5. Comparison of Cancer Mortality Rates of 21 Central Cities and Suburbs, 1950-59 and 1970-75

| | Central Cities/Suburbs | | | |
| | White Male | | White Female | |
SMSA	1950–59	1970–75	1950–59	1970–75
1. Atlanta	1.29	1.01	1.20	1.04
2. Baltimore	1.27	1.22	1.18	1.12
3. Boston	1.22	1.14	1.08	1.09
4. Buffalo	1.11	1.02	1.09	1.00
5. Chicago	1.20	1.08	1.13	1.09
6. Cincinnati	1.11	1.02	1.07	1.04
7. Cleveland	1.33	1.10	1.06	1.02
8. Dallas	1.21	1.19	1.14	1.12
9. Detroit	1.17	1.08	1.04	1.00
10. Kansas City	1.31	1.16	1.01	1.12
11. Milwaukee	1.27	1.08	1.08	1.04
12. Minneapolis/St. Paul	1.35	1.14	1.14	1.00
13. New York	1.14	1.01	1.15	0.99
14. Newark	1.10	0.97	1.04	1.04
15. Paterson/Passaic/Clifton	1.05	1.02	0.98	1.02
16. Philadelphia	1.20	1.15	1.12	1.04
17. Pittsburgh	1.12	1.13	1.06	1.09
18. St. Louis	1.25	1.30	1.15	1.06
19. San Francisco	1.22	1.17	1.12	1.10
20. Seattle	1.18	1.03	1.07	1.01
21. Washington, D.C.	1.25	1.00	1.07	1.12
Highest	1.35	1.30	1.20	1.12
Median	1.21	1.08	1.08	1.04
Lowest	1.05	0.97	0.98	0.99

Minneapolis/St. Paul SMSA had a total white male cancer mortality rate 35 percent higher than the suburban counties. At the other end, the difference between the central cities and suburbs of the Paterson/Passaic/Clifton SMSA was only 5 percent.

Twenty of the 21 SMSA central cities had higher white female cancer mortality rates than their suburbs during 1950–59. The median was 8 percent higher with a range of 20 percent higher (Atlanta) to 2 percent lower (Paterson/Passaic/Clifton).

The Paterson/Passaic/Clifton exception does not weaken the supporting evidence. The central city county is Passaic County, which contains three cities. The suburb is in Bergen County, which, although it contains no large cities, is as much or more a suburb of New York City than of Passaic County. Indeed, for a time, Bergen County was considered part of the New York, New York region. A further look into the rates of this SMSA disclosed that the major reason that the suburban female mortality rate was higher than the central city rate was due to female breast cancer. The white 1950–59 female breast cancer mortality rate in Passaic County, the central city, was 29.3/100,000; it was 33.6/100,000 in far more affluent suburban Bergen County. This is not an unexpected result because female breast cancer mortality rates are reported to increase with affluence (Blot, 1977). Indeed, cancer of the female reproductive organs is the major reason why convergence between central cities and suburbs has been less among females than among males. Cancer of the breast, cervix, uteri, and corpus uteri do not exhibit the strong trend toward relative increases in the suburbs in comparison to central cities.

During 1970–75, the central cities continued to have higher cancer mortality rates than the suburbs. However, as expected, the gap between the central cities and the suburbs narrowed. The median white male and white female central city rates were respectively 21 and 8 percent higher than their suburban counterparts during 1950–59. By 1970–75 the difference narrowed to 8 percent for males and 4 percent for females.

Nineteen of the 21 white male, central city/suburb ratios exceeded 1.0 compared to all 21 ratios during 1950–59. Newark and Washington, D.C. were the exceptions; their ratios were 1.0 or below 1.0. During the study period, two (St. Louis and Pittsburgh) of the 21 central city/suburb ratios increased.

With respect to white females, 17 of the 21 central city/suburb ratios exceeded 1.0 during 1970–75, compared to 20 of 21 ratios two decades earlier. New York, Buffalo, Detroit, and Minneapolis/St. Paul were the exceptions. Sixteen of the 21 ratios decreased for white females, leading to a narrowing of the central city/suburb gap. Boston, Kansas City, Newark, Pittsburgh, and Washington, D.C. were the exceptions.

In conclusion, despite the data limitations discussed above, it is clear that although high cancer mortality rates were primarily characteristic of large central cities through the mid-1950's, they are now characteristic of entire metropolitan regions. Male rates are generally increasing, but more so in suburbs than in central cities, and female rates are decreasing, but usually less so in suburbs than in central cities.

This demonstrates that the distinctive geographical profiles of cancer have become blurred. The major differences observed during the early 1950's between the most urbanized and least urbanized counties, between urbanized regions of the Northeast, Northcentral, South, and West and between central cities and suburbs have been disappearing to be replaced by increasingly homogeneous national white male, white female, and probably nonwhite male, and nonwhite female profiles.

There appears to have been pronounced changes in the cancer mortality profiles of white males in the major central cities of the United States. Accordingly, the cancer profiles of white males in the 21 major central cities are presented in greater detail with respect to changes in white male cancer mortality rates during the period 1950–59 through 1970–75.

The 21 central cities may be divided into four groups (below).

Group 1: Decreasing rates	Group 2: Increase less than 8%	Group 3: Increase 8 to 16%	Group 4: Increase greater than 16%
Milwaukee	Boston	Atlanta	Baltimore
New York City	Buffalo	Pittsburgh	Cincinnati
Newark	Chicago	Minneapolis/	Dallas
Paterson/Passaic/	Cleveland	St. Paul	Kansas City
Clifton	Detroit	San Francisco	St. Louis
Washington, D.C.	Philadelphia	Seattle	

Five of the 21 cities exhibited decreasing rates. Four of these five Group 1 cities are along the eastern seaboard. The fifth, Milwaukee, borders the Great Lakes. The six Group 2 central cities with slight increases are all along the Great Lakes or the eastern seaboard. Three of the five Group 3 cities are in the West, one is on the Northern Plains. The group with the greatest increases (Group 4) is primarily in the Midwest and South. Overall, higher rates of increase are associated with Midwestern, Western, and Southern cities and decreasing and stable rates with the eastern seaboard and Great Lakes cities.

This finding is not surprising in light of the tendency reported in Chapter 4 toward increasing homogeneity between, on the one hand, the Northeastern and Northcentral and, on the other hand, the Southern and Western strongly urban areas. What is surprising is the range of differences between the Group 1 and the Group 4 cities. Rates in the Group 1 cities decreased, whereas rates in the Group 4 centers increased more than the United States as a whole.

A two-pronged study was made in order to try to understand what cities with such diverse locations as Milwaukee and Newark (Group 1) and Baltimore and Dallas (Group 4) might have in common. The first prong focused on types of cancers responsible for the range of differences, the second on reasons for the differences.

The first prong revealed that the cities with decreasing white male rates exhibited substantial decreases in digestive system and relatively slight increases in respiratory cancer mortality rates (Figure 6-3). The converse was true for the five cities that manifested the greates increases. New York City, Milwaukee, Paterson/Passaic/Clifton, and Newark all began the 1980's with digestive tract cancer mortality rates higher than 90/100,000, about 50 percent higher than the national average. Two decades later, their rates had dropped to between 55 and 70/100,000.

These four Group 1 cities stood in obvious contrast to the five Group 4 cities, all of which began with lower digestive tract cancer mortality rates and exhibited much smaller decreases in digestive tract cancers. Among the Group 1 cities, only Washington, D.C. began with a relatively low digestive tract rate and during the next two decades manifested a relatively small decrease in those rates.

The second component of the difference between the Group 1 and Group 4 cities is cancer of the respiratory tract. All ten cities began the 1980's with white male mortality rates between 30 and 45/100,000. However, two decades later, the five Group 1 cities had respiratory cancer mortality rates between 55 and 65/100,000, compared to 65 to 95/100,000 for the Group 4 cities (Figure 6-3).

The first prong found substantial differences in the rates of change of white male cancer mortality in central cities, and so the second prong of the analysis suggests and tests explanatory hypotheses. Most of these have already been presented elsewhere in the book.

One set of hypotheses focuses on population migration. Increasing cancer mortality rates in some central cities may be due to immigration of persons who formerly lived in areas with a higher risk of contracting cancer. Decreasing cancer mortality rates in other cities may be due in part to the emigration of the cities' elderly to the suburbs or to sunbelt cities. Another plausible component of the migration factor is the migration of white, Hispanic persons born in Latin America to some American cities. The Hispanic population tends to have lower cancer mortality rates than the non-Hispanic city population. If this population has concentrated in the Northeastern cities, its presence could help explain the decreasing cancer rates in those cities.

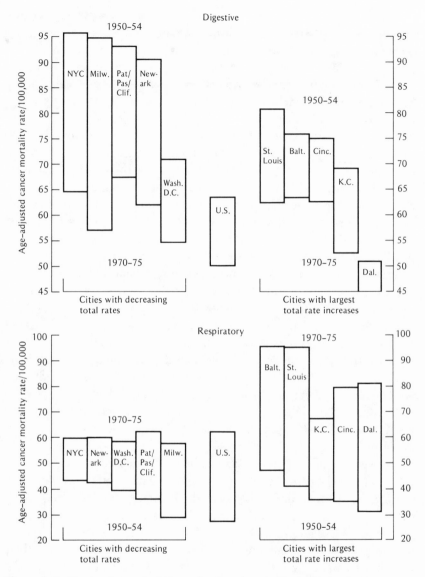

Figure 6–3. Comparison of white male digestive and respiratory system cancer mortality rates in central cities with decreasing and increasing rates, 1950–54 and 1970–75

A second set of hypotheses is labeled the changing geographical distribution of risk. The trend toward similar rates of industrialization and of smoking habits across the United States (see Chapter 4) would help account for the rapid increase of respiratory-tract cancer mortality rates in the Group 4 cities. A decrease in the number of European-born Americans might explain the marked decrease of digestive tract cancer mortality in the Group 1 cities.

The last set of hypotheses focuses on the size, shape, and content of the central cities. As previously pointed out, central cities in this study are in counties that contain populous central cities. For some of the 21 central cities, the counties may fit the classical mold of the high density city, others may include large suburban and exurban areas within their boundaries. In short, part of the reason why the Group 1 and Group 4 cities have changed at different rates may be due to differences in what constitutes the working definition of central city for this study.

Testing the hypotheses is difficult because of data limitations. Nevertheless, tests using the available data lead to some intriguing results. First, the white foreign-born link to digestive tract cancer is quite suggestive. Specifically, New York City, Milwaukee, Passaic (Paterson/Passaic/Clifton), and Essex (Newark) counties had much higher while male digestive tract cancer mortality rates than the remaining six central city counties during 1950–54. In 1950, between 10 and 24 percent of their white population was foreign-born. In comparison, the remaining six cities that had digestive tract rates between 20 and 50 percent less than the above four had a white population that was only between 2 and 7 percent foreign-born.

The very significant decreases in white male digestive tract cancer mortality rates during the study period coincide with pronounced decreases in the proportion of foreign-born among the white population, especially in New York, Milwaukee, Passaic, and Essex counties.

The second suggestive link is between cigarette smoking and respiratory cancer. Before presenting the results it must be repeated that there are no systematically collected county smoking data available. Accordingly, in order to compare Group 1 and 4 counties, two assumptions had to be made. First, trends in the states for which data are available must be assumed to parallel trends in the central city counties. Second, as discussed in Chapter 4, out-of-state purchases of cigarettes are increasingly common. Out-of-state purchases, however, must be assumed to be relatively unimportant for the eight states of reference. If these assumptions are accepted, the results imply that cigarette smoking has increased much more rapidly in the Group 4 than in the Group 1 cities. Specifically in 1960, the first date for which cigarette smoking data are available for all eight states and the District of Columbia, the three states (New Jersey, New York, and Wisconsin) and the District of Columbia ranked first, second, third, and seventh out of nine in per capita cigarette sales. The Group 4 states ranked fourth, fifth, sixth, eighth, and ninth. Two decades later, in 1980, the rankings were quite different. Per capita sales in the District of Columbia, New Jersey, and New York had decreased. Sales in the Group 4 states all increased. The final result was that the states containing the Group 1 cities dropped from first, second, third, and seventh to fifth, sixth, eighth, and ninth in per capita cigarette sales. The Group 4 states increased from fourth, fifth, sixth, eighth, and ninth to first, second, third, fourth, and seventh. The obvious shortcomings of the data may invalidate the results, and they certainly weaken them. Nevertheless, the results were presented because they are so suggestive of an often-stated relationship between cigarette smoking and respiratory tract cancer.

The other hypotheses did not yield strong results. If rates in the Group 1 cities are declining because of migration, then net out-migration should be pronounced, and the white population should be decreasing. If rates in the Group 4 cities are increasing because of the net in-migration of persons from areas of higher risk, then net migration rates, especially of whites, should be positive. The expected was not consistently found. Among the Group 1 cities, those in Passaic County, which recorded a marked decrease in digestive tract cancer mortality, recorded a net in-migration of people and an increased white population between 1950 and 1970. Milwaukee, another Group 1 city, also increased in white population during the two decades and recorded a net increase in migration during the 1950's. Of the five Group 1 cities, only New York City, Essex County, and Washington, D.C. exhibited an out-migration of the white population that might be responsible for their decreasing digestive tract cancer mortality rates.

The results for the Group 4 cities were no more supportive of the migration hypothesis than they were for the Group 1 cities. Dallas clearly experienced net in-migration, and along with Dallas, Kansas City and Cincinnati also showed an increase in the number of whites between 1950 and 1970. However, with the exception of Baltimore, few of the migrants to the Group 4 cities came from the most urban states. Most, as noted in Chapter 4, came from the same state and its neighboring states.

In addition, two of the five Group 4 cities, St. Louis and Baltimore, experienced substantial net out-migration and decreases in their white populations. Hispanic in-migration does not offer a satisfactory explanation because, with the exception of New York City and Dallas, it is too small and too young to account for the major shifts observed in the ten cities. Although migration hypotheses do not offer a consistent answer, they may help explain the central city/suburban results for Washington, D.C. and St. Louis. The differences between the total white male cancer mortality rates in Washington, D.C. and its suburbs dropped dramatically from 1.25 during 1950–59 to 1.00 during 1970–75. The massive suburbanization of the capital district probably played an important role. In strong contrast to Washington, D.C., the gap between the central city and the suburbs of St. Louis increased from 1.25 to 1.30. Massive suburbanization of the white population was also true of St. Louis. Perhaps, the population that remained in the city and the new population were at higher risk than the much larger white population that moved to the suburbs.

The spread of industry and higher occupational risk could play an important role, particularly in respiratory tract cancer. However, although recognizing the fact that the results may be undermined by the use of all production jobs as an indicator of occupational risk, a review of the changing distribution of production employment failed to provide a consistent answer. Among the ten counties, Passaic, Milwaukee, and Essex ranked first, second, and third in indicators of manufacturing production employment when the study period began. However, New York and Washington, D.C. ranked eighth and tenth. Yet among the five Group 1 cities, New York City exhibited higher white male respiratory tract cancer mortality rates during 1950–54 than the other four Group 1 cities. Washington, D.C.'s rate was higher than all but New York City's and Newark's.

In short, the assumption that production employment is a reasonable indicator of occupational exposure to carcinogenic substances leads to rejection of the occupational exposure hypotheses for these ten central cities because there is no obvious association between respiratory tract cancer mortality rates and the role of production employment in the local economy. Suffice it to say that the occupational hypothesis might be more convincing if better data were available.

AMENITY RETIREMENT AREAS: THE DISTANT SUBURBS

The first part of this chapter focused on conventional suburbs, places adjacent to cities where many people moved when their families expanded, and required more space or when they became sufficiently affluent to afford more space. Moving to the suburbs, however, was not the final move for many of the more affluent. Many have moved to retirement areas and, in effect, have created new suburbs. In this section, we examine the extent to which the cancer profiles of amenity retirement areas look like those of the suburbs as opposed to the rural areas in which they are situated.

It is hard to imagine the elderly at the front of an urban to rural migration stream. Elderly migration rates are only about one-half the national average (U.S. Bureau of the Census, 1970, 1980a). An increase in the proportion of elderly in a community is almost always due to the out-migration of younger people and the aging-in-place of the elderly. The combination of migration of the young and aging-in-place resulted, in 1950, in elderly concentrations in northern New England (Maine, New Hampshire, and Vermont) and the Midwest and adjacent Great Plains (Indiana, Iowa, Kansas, Missouri, and Nebraska). In 1980, the highest concentrations were in the Midwest and Great Plains and some Southern states; notably, Arkansas, Florida, and Oklahoma had much larger proportions of elderly in 1980 than they did in 1950.

The increasing size of the elderly population and its growing economic and political power have made the hard-to-imagine occur: the resettlement of the urban elderly in large numbers in states like Arkansas, Florida, and Oklahoma. This large increase from 3.1 million elderly in 1900 to more than 25 million elderly in 1980 (from 4.1 to 11.3 percent of the American population) is one factor in the growing demographic and political strength of this population (U.S. Bureau of the Census, 1976, 1981). Elderly incomes are only about one-half the national average. However, Social Security payments and pension plans have given many elderly an opportunity to relocate (U.S. Bureau of the Census, 1980b). During the period 1950 to 1979, the number of beneficiaries of Social Security programs tripled and the average payment per person (in constant dollars) increased 80 percent. At the same time, the number of people benefiting from private pension plans and deferred profit-sharing plans increased more than 15 times, and the dollars they received more than 40 times.

The procedure was to determine if formerly rural areas have become amenity retirement areas (AMR) for the elderly and then to determine if they have experienced marked changes in the number and types of cancer deaths. The first step of demonstrating that "specialized amenity retirement areas" were formed in rural areas was taken by Graff and Wiseman (1978). Defining persons age 65 and older as the elderly, and the county as the unit of analysis, the following three criteria were applied to all counties in the United States: (1) at least 15 percent of the population is elderly (one and one-half the national proportion of elderly), (2) an increase between 1950 and 1970 in the proportion of elderly persons of at least 5.4 percent (three times the national increase), and (3) a net increase in migrants, which eliminates counties that have had an increase in the elderly due to the out-migration of younger people.

Employing the three criteria set forth above, Graff and Wiseman found 94 counties in 17 states that we will call amenity retirement counties. The preponderance are in the South. Sixty-four of the counties (68 percent) are in four Southern states: Arkansas, 19; Florida, 17; Oklahoma, 8; and Texas, 20. Thirteen of the remaining 30 are in Michigan, 5; Missouri, 4; and Wisconsin, 4.

As a whole, 68 percent of the people living in the 94 amenity retirement counties were classified as urban in 1970. The 68 percent figure is misleading insofar as 68 percent implies a uniformly highly urbanized population in all or even a majority of the AMR counties. Almost 85 percent of the urban population in the AMRs is in one state: Florida. Of the population of the 17 Florida counties, 83 percent was classified as urban in 1970 compared to only 30 percent in Arkansas, 30 percent in Oklahoma, and 38 percent in Texas. Few amenity retirement counties are in metropolitan areas. Only one city of more than 100,000 people, St. Petersburg, Florida is in an AMR, and only five counties (three in Florida) are in Standard Metropolitan Statistical Areas. About 25 percent of the AMR counties were more than one-half urban in 1970, whereas three-quarters were rural. About one-third of the counties had no urban population in 1970. In short, the amenity retirement counties are largely rural counties.

The second step was to determine changing cancer mortality patterns among the elderly in these counties while controlling for national trends in cancer and for urbanization. Age-adjusted rates were calculated for the white male and white female populations 65 and older using the methods described in Chapter 2. The rates are divided by the national rate so that an index of 100 means that the rate of the area equals the U.S. rate.

Six specific types of cancer, by site, were chosen to see if the cancer mortality profiles of the amenity retirement counties have been moving toward the national pattern; as shown below.

Cancer Sites (ICD, 7th revision)	Description	Population
153, 154	Large intestine and rectum	White male; white female
162, 163	Trachea, bronchus, and lung specified as primary; and lung and bronchus unspecified as to whether primary or secondary	White male
170	Breast	White female
175	Ovary, Fallopian tube, and broad ligament	White female
181	Bladder and other urinary organs	White male; white female
191	Skin, not including melanoma	White male

As demonstrated in earlier chapters, five of the six cancer categories (skin is the exception) have had considerably higher urban than rural cancer mortality rates. Differences between the urban and rural areas have been declining. Accordingly, tracing the changes in the mortality rates of these types of cancer should show if and how rapidly the amenity retirement counties have been changing in their disease profiles.

Results for each of the six types of cancer are presented for an aggregate of all of the amenity retirement counties and, as controls, the United States and the strongly urban, moderately urban, and rural aggregates. In addition, separate results are presented for the four states that had at least eight AMR counties and a population of 125,000 (Arkansas, Florida, Oklahoma, and Texas) and for the three types of cancer with age-adjusted elderly cancer mortality rates of at least 100/100,000 (white male lung, white male and female large intestine and rectum, female breast). Detailed presentations are limited to these types because the combination of less populous states and low rate diseases leads to very high confidence bands around the rates and less confidence in comparisons. To aid comparisons with the United States, significance levels of $p \leq .01$ and $p \leq .05$ are presented for the amenity retirement counties.

As expected, the AMR's experienced a major increase in the number of elderly cancer deaths from the early 1950's to the mid-1970's. Elderly cancer deaths increased 87 percent in the United States, ranging from a high of 91 percent in the strongly urban counties to a low of 76 percent in the rural counties. In comparison the number of cancer deaths among the elderly in the AMR's increased 375 percent with the most pronounced increase for cancer of the trachea, bronchus, and lung: 1184 percent. In summary, when measured by the number of cancer deaths the amenity retirement counties have had an increase that dwarfs the increases experienced in the United States.

The pattern of elderly cancer deaths in the AMR counties has clearly moved toward an urban profile, particularly in the less urbanized amenity retirement counties. Beginning with white male trachea, bronchus and lung cancer, age-adjusted rates for the elderly in the continental area part of the United States were 27 percent higher than the rates in the amenity retirement counties in 1950–54 (Table 6–6). In 1970–75, the difference had shrunk to 12 percent. The less urbanized AMR counties in Arkansas, Oklahoma, and Texas show a more rapid increase toward the national rate than even the rural counties of the United States, with Florida, Oklahoma, and Texas exhibiting rates similar to the moderately

Table 6–6. Comparison of White Population, Elderly Cancer Mortality Rates in Amenity Retirement Areas of the United States, the United States, and Strongly Urban, Moderately Urban, and Rural Areas, 1950–54 and 1970–75

	Ratio of Area Rate/U.S.A. Rate							
	Amenity Retirement Areas[a]							
Type of Cancer	Aggregate[a]		Arkansas[b]		Florida[c]		Oklahoma[d]	
	1950–54	1970–75	1950–54	1970–75	1950–54	1970–75	1950–54	1970–75
Male lung	.79[f]	.89[f]	.50[f]	.75[f]	.88[g]	.95[f]	.57[f]	1.01
Female breast	.69[f]	.84[f]	.52[f]	.81[f]	.73[f]	.86[f]	.48[f]	.59[f]
Male large intestine, rectum	.67[f]	.88[f]	.73[f]	.91[f]	.32[f]	.69[f]	.47[f]	.65[f]
Female large intestine, rectum	.85[f]	.88[f]	.77[f]	.85[g]	.76[f]	.87[f]	.75[f]	.76[f]
Ovary	.72[f]	.88[f]	__[h]	__[h]	__[h]	__[h]	__[h]	__[h]
Male bladder	.73[f]	.85[f]	__[h]	__[h]	__[h]	__[h]	__[h]	__[h]
Female bladder	.81[f]	.94	__[h]	__[h]	__[h]	__[h]	__[h]	__[h]
Male skin (excl. melanoma)	1.15	1.03	__[h]	__[h]	__[h]	__[h]	__[h]	__[h]

[a] 94 counties in 17 states.
[b] 19 counties.
[c] 17 counties.
[d] 8 counties.
[e] 20 counties.
[f] $p \leq .01$, presented for amenity retirement areas.
[g] $p \leq .05$, presented for amenity retirement areas.
[h] Not presented because of wide confidence limits.

urban (50–74.9 percent urban) aggregate of the United States in 1970–75.

The trend toward an urban profile is equally apparent for female breast and male and female large intestine and rectum cancer (Table 6–6). Rates in the United States for these three types of cancer decreased from 44, 49, and 18 percent higher than the AMRs during 1950–54 to 19, 14, and 14 percent higher in 1970–75. Although rates in the amenity retirement counties were usually below the rural United States aggregate, they have moved toward the urban profile more rapidly than the rural aggregate of the United States, particularly in the less urbanized Arkansas, Oklahoma, and Texas AMR's.

The AMR's have also narrowed urban/rural differences more rapidly than the rural aggregate of the United States for bladder, ovarian, and non-melanoma skin cancer (Table 6–6). Skin cancer, excluding melanoma, is particularly interesting because it represents the example of a disease for which rural rates have been higher than the national rate. Although the white male, skin cancer mortality rate of the rural aggregate of the United States decreased from 21 percent higher to 19 percent

Ratio of Area Rate/U.S.A. Rate

Areas of the United States

Texas[e]		Strongly Urban		Moderately Urban		Rural	
1950–54	1970–75	1950–54	1970–75	1950–54	1970–75	1950–54	1970–75
.66[f]	.94	1.27	1.07	.84	.98	.66	.89
.52[f]	.65[f]	1.11	1.07	.91	.95	.85	.87
.46[f]	.66[f]	1.20	1.11	.89	.95	.76	.83
.61[f]	.75[f]	1.08	1.04	.92	.97	.90	.92
—[h]	[h]	1.12	1.04	.94	.98	.82	.93
—[h]	[h]	1.25	1.11	.88	.94	.70	.83
—[h]	[h]	1.09	1.05	.95	.98	.86	.91
—[h]	[h]	.84	.87	1.04	1.07	1.21	1.19

higher than the national rate, the rate in the amenity retirement county aggregate decreased from 15 percent higher to only 3 percent higher during the study period. Although noteworthy, this last change in the AMR's must be cautiously interpreted because the white male skin cancer mortality rate in the AMR's is based on 95 records during 1950–54 and 142 records during 1970–75, which produces rates that one cannot be 95 percent confident are significantly different from the national rate.

In conclusion, as expected, the amenity retirement counties have not only experienced a large, perhaps unprecedented, increase in the number of elderly cancer deaths, but in addition, what was a rural cancer profile when the study period began has become increasingly a suburban profile occurring in ostensibly rural areas.

SUMMARY

This chapter focuses on three questions, which will be repeated and briefly answered.

1. How do cancer mortality rates in central cities and suburbs compare with those of less urbanized areas of the United States?

Data from the central cities and suburbs of 22 of the 25 most populated metropolitan areas yielded the expected urban/rural differences. During 1950–54, the central city counties had total white male rates that were 51, 26, and 17 percent higher than the rural counties of the United States, the continental United States, and the suburban counties, respectively. The analogous comparisons for white females were 26, 15, and 11 percent. Particularly marked differences were observed for the respiratory, digestive, and urinary systems and female breast cancer.

By 1970–75, these differences had decreased. Differences between total white male rates in central cities, rural counties, the continental United States, and the suburbs dropped from 51, 26, and 17 percent to 19, 10, and 5 percent. Changes in white female differences were less marked; from 26, 15, and 11 to 18, 10, and 3 percent.

2. How do central city and suburban cancer mortality profiles differ?

During the 1950's, every one of the 21 central cities had higher white male total cancer mortality rates than its suburbs. The same was true in 20 of the 21 comparisons of white female total cancer mortality rates. There were marked differences for the respiratory and digestive systems. Differences had substantially narrowed two decades later. The median white male and white female central city rates were 21 and 8 percent higher than the suburban rates during 1950–59. By 1970–75, the median difference had decreased to 8 percent for males and 4 percent for females.

In particular, white male cancer mortality rates in central cities were analyzed because of the range of differences in changes during the study period. One set of central cities (Group 1) manifested decreasing rates [New York City, Essex (Newark), Passaic (Paterson/Passaic/Clifton), Milwaukee, and Washington, D.C.], whereas a second set (Group 4) manifested greater increases than the continental United States as a whole [St. Louis, Baltimore, Kansas City, Hamilton County (Cincinnati), and Dallas]. The Group 1 cities manifested decreasing total, white male cancer mortality rates because of pronounced decreases in digestive system cancer and relatively small increases in respiratory system cancer mortality rates. In strong contrast, the rates of Group 4 cities increased in total white male cancer mortality; they were higher than the national increase because of relatively small decreases in digestive system combined with major increases in respiratory cancer mortality rates.

The two best clues to an explanation of the above are the marked decrease in the white foreign-born population in the Group 1 cities, especially New York City, Paterson/Passaic/Clifton (Passaic County), Newark (Essex County) and Milwaukee. Albeit weak because of the necessity of applying state data to cities, there is evidence to suggest that cigarette use has increased substantially in the Group 4 cities relative to cigarette use in the Group 1 cities.

3. Do the cancer mortality profiles of amenity retirement areas reflect the rural areas in which they are situated or the cities and suburbs from where much of their population came?

As a result of their increasing numbers and growing economic and political power, many elderly have created new suburban amenity retirement areas, especially in Arkansas, Florida, Oklahoma, and Texas. These amenity retirement areas have experienced a marked increase in the number of cancer deaths, and their cancer mortality profiles increasingly resemble those of urban America.

REFERENCES

Berry, B. 1973. *The Human Consequences of Urbanization*. New York: St. Martin's Press.

Blot, W., Fraumeni, J., and Stone, B. 1977. "Geographic patterns of breast cancer in the United States." *J. Nat. Can. Inst.* 59: 1407–1411.

Brunn, S., and Wheeler, J. 1980. *The American Metropolitan System: Present and Future*. New York: John Wiley.

Graff, T., Wiseman, R. 1978. "Changing concentrations of older Americans," *Geog. Rev.*, vol. 68, pp. 379–393.

Handlin, O., and Burchard, J., eds. 1966. *The Historian and the City*. Cambridge: Harvard Univ. Press.

Masotti, O., and Hadden, J. 1973. *The Urbanization of the Suburbs*, vol. 7, Urban Affairs Annual Review. Beverly Hills: Sage, 1973.

Wells, H.G. 1902. *Anticipations. The Reaction of Mechanical and Scientific Progress on Human Life and Thought*. London: Harper and Row.

Wirth, L., "Urbanism as a way of life." 1938. *Amer. J. Sociol.*, XLIV: 1–24.

U.S. Bureau of the Census. 1970. *1970 Census of Population, Subject Reports, Mobility for States and the Nation*. Washington, D.C.: U.S.G.P.O.

U.S. Bureau of the Census. 1980a. *Current Population Reports, Population Characteristics, Geographical Mobility, March 1975 to March 1979*. Washington, D.C.: U.S.G.P.O.

U.S. Bureau of the Census. 1976. *Current Population Reports, Special Studies, Demographic Aspects of Aging in the United States*, Washington, D.C.: U.S.G.P.O.

U.S. Bureau of the Census. 1981. *Age, Sex, Race, and Spanish Origin of the Population by Regions, Divisions, and States: 1980*. Washington, D.C.: U.S.G.P.O.

U.S. Bureau of the Census. 1980b. *Statistical Abstract of the United States, 1980*. Washington, D.C.: U.S.G.P.O.

7 Summary and Discussion

The United States has lower cancer mortality rates than most Western, urban/industrial nations and higher rates than most non-Western and less industrialized nations. During the period 1950–75, cancer of the lung, pancreas, connective tissue, and central nervous system and multiple myeloma, melanoma, and non-Hodgkin's lymphoma have become far more important causes of mortality in the United States. Cancer of the stomach, rectum, liver, cervix uteri, corpus uteri, bone, non-melanoma skin, and lip has become much less important during the same period. A review of the limited time series data for different nations indicates that most of these trends are common to all Western, urban/industrial nations.

In 1930, white females had the highest and nonwhite males the lowest cancer mortality rates in the United States. These positions reversed by 1970–75. Male rates have been increasing since at least 1930, while female rates have been declining and, more recently, stabilizing. Trends among white males, nonwhite males, white females, and nonwhite females are similar for nearly every type of cancer. But nonwhite males have almost always had the largest increase or smallest decrease, whereas white females have almost always had the smallest increase or the largest decrease.

The elderly are primarily responsible for trends in cancer mortality. They exhibit the highest rates and almost always the most pronounced changes.

It has been consistently reported that respiratory (especially lung), digestive (especially intestine), urinary (especially bladder), and female breast cancer mortality rates are much higher in urban than in rural areas. In the effort to test the reports described in this book, all counties in the continental United States were divided into three groups: strongly urban (greater than or equal to 75 percent urban); moderately urban (50–74.9 percent urban); and rural (less than 50 percent urban). Comparisons among the three groups of counties for the years 1950–54 showed the expected urban/rural differences for every one of the expected sites. The urban/rural ratios for total white male and white female cancer were 1.35 and 1.16, respectively. Seven ratios exceeded 1.5: male esophagus; male larynx; male bladder; male tongue/mouth; male and female rectum; and male lung. Eight ratios were less than 1.5, but greater than 1.3: male large intestine; male pancreas; male and female lymphoma; male kidney; female breast; ovary; and female bladder.

By 1970–75, everyone of these ratios had fallen. Only the male tongue/mouth ratio continued to exceed 1.5. The urban/rural ratios for total cancer fell from 1.35 to 1.12 for white males and from 1.16 to 1.12 for white females.

Nonwhite urban/rural comparisons are limited by a smaller population base and by the widely differing cancer rates and geographical distributions of major nonwhite American populations: blacks, Chinese, Japanese, and American Indians. With these caveats in mind, limited comparisons showed even stronger urban/rural differences for nonwhites than for whites and a decrease in these differences during the study period.

An analysis of cancer mortality data for 52 nations suggested the view that as urbanization increases, mortality rates increase for cancer of the esophagus, intestines, lung, the lymphatic system, male rectum, and female breast cancer. Data from Poland, Norway, Canada, and England and Wales showed urban/rural differences for the expected sites. In addition, the Polish data revealed declining urban/rural cancer gradients.

When the study period began, there were strong associations between the level of urbanization and per capita cigarette smoking and alcohol consumption at the state scale in the United States. In addition, the most urbanized states had a much higher proportion of white foreign-born inhabitants, as well as affluent people, and production jobs, than the rural states. The most rural states had a larger proportion of poor people than the most urban states. It is argued that the geographical distribution of these and other risk indicators, along with differences in medical practices perhaps, were largely responsible for urban/rural differences in cancer mortality during the 1950's and earlier.

By the 1970's, there was little difference between the most urban and least urban states in per capita cigarette and alcohol sales and of foreign-born, white Americans and production workers. In addition, relative poverty was more characteristic of urban areas than two decades earlier. Along with population migration and changes in medical practices, the dramatic geographical shift of risk factors should explain the trend toward a homogeneous geography of cancer in the United States.

In order to assess the role of population migration, the migratory stream to the eight most rural states was analyzed. At least in the cases of Mississippi, North Carolina, South Carolina, and West Virginia, interstate migration probably plays an insignificant role in the substantial increases in white male cancer mortality rates.

The Northeastern and to a lesser extent the Northcentral metropolises had higher cancer mortality rates than their Southern and Western counterparts when the study period began, especially for the digestive organs. Differences between the four regions markedly narrowed during the 1950–75 period. Perhaps the most intriguing example is white female respiratory cancer. During 1950–54, the highest rates were found in the Northeastern and Northcentral urban areas. Two decades later, the highest rates were in the Western and Southern metropolises.

During 1950–69, the most urbanized counties of the continental United States were found to have significantly higher child and teenage cancer mortality rates than the nation as a whole. Among the approximately 300 most urbanized counties of the United States, the highest child and teenage cancer mortality rates tended to be in the New York/ New Jersey region and northern California. Some New England and Great Lakes Region counties exhibited high rates of particular types of child and teenage cancer.

During 1950–54, central city counties had much higher cancer mortality rates than suburban counties and the remainder of the United States for almost every type of cancer. Two decades later, these differences had markedly narrowed. Specifically, differences between central cities on the one hand, and, on the other, rural counties, the United States as a whole, and suburban counties in total, white male cancer mortality decreased, respectively, from 51, 26, and 17 percent during 1950–54 to 19, 10, and 5 percent during 1970–75. White female differences decreased from 26, 15, and 11 percent to 18, 10, and 3 percent.

Almost every one of the central cities had higher cancer mortality rates than its suburbs during the 1950's and still had higher rates two decades later. However, the city/suburban differences decreased sharply during these two and one-half decades.

A special analysis was made of white male cancer mortality rates in central cities because of the range of differences in change experienced by these cities. New York City, Essex County (Newark), Passaic County

(Paterson/Passaic/Clifton), and Milwaukee had decreasing total rates because of decreases in digestive cancer and relatively small increases in respiratory cancer. These four cities contrast sharply with Dallas, Kansas City, St. Louis, Baltimore, and Cincinnati, each of which had increases in the total, white male cancer mortality rate greater than the national increase due to small decreases in digestive cancer combined with pronounced increases in respiratory cancer. Further investigation suggested that the decreases in digestive cancer paralleled decreases in the white foreign-born population of these central cities. The increase in respiratory cancer in some of the cities and not in others may have resulted from changes in cigarette use. However, the data that support this hypothesis were at the state scale and therefore may not accurately characterize the cities.

A special study of amenity retirement areas showed a marked increase in the number of elderly cancer deaths during 1950–75, and the cancer profiles of the amenity retirement areas increasingly resemble those of cities and suburbs rather than the rural areas in which they are situated.

An Avenue for Future Research

The main research path indicated by this book is clear: the trend toward homogeneity. It is an important trend to study because of its policy implications. The urban areas can serve as a blueprint for the suburban and rural areas with respect to cancer. If geographical differences are disappearing for other chronic diseases, health planners will now be better able to forecast personnel and equipment needs than they were in the past.

Two types of research are suggested. One type is to study trends toward homogeneity for other diseases in the United States and for cancer and other chronic diseases in other nations. Testing for homogeneity requires a data base parallel to the cancer death base used in this volume. Other nations, notably the United Kingdom, Canada, and the Federal Republic of Germany, have long-standing cancer records. Still others are developing data bases.

The second type of research is to determine the origins of the trend documented in these pages. Each plausible major factor should be investigated. For example, little is known about the variation of medical practices by region. Sample studies of medical practices, controlled for regional population size, degree of urbanization, by region, and by demographic characteristics of the population, would help determine the role of variations in medical practices in biasing mortality and morbidity records. Differences in mortality rates by length of stay in a region and by region of origin, controlled for race, age, and sex, would help quantify the role of inter- and intraregional migration in changing the geography of chronic diseases in the United States and other nations. In order to isolate the role of disease competition, a profile of diseases with temporal and spatial dimensions should be constructed. For example, it would be interesting to know if areas with infamous labels such as New Jersey's "cancer alley" owe their reputations at least partially to lower death rates from competitive diseases.

Perhaps the most important thing to measure is the changing distrubition of etiological factors, whether through censuses or more likely sample surveys. Without data on cigarette smoking, alcohol intake, dietary practices, ethnicity, occupation, and ambient environmental and other possible etiological factors, we can only speculate about the meaning of mortality and morbidity trends.

Research on unusual findings should not be neglected. The increase in nonwhite male cancer mortality rates is so remarkable that it warrants immediate attention. Perhaps, as some have suggested, this dramatic increase is an artifact of initial undercounting and present low survival rates due in part to poverty and cultural factors that preclude early diagnosis. On the other hand, black males may have occupations, residences, or life-styles that increase their risk of contracting cancer. Another unusual finding is that white female, respiratory cancer mortality rates are rising more rapidly in some urbanized areas than in rural areas, perhaps due to different cigarette smoking rates among females or different occupational and air pollution exposures in urban than in rural areas. This bears study, as do rural-to-urban migrant groups such as Hispanics and Asians who have come from places with cancer mortality profiles that strongly contrast with the U.S. norm.

1 Supporting Tables

This appendix presents 50 tables supporting the graphics in Chapters 3 to 6. Each table is cited in the text. Tables A1–1 through A1–13 go with Chapter 3, Tables A1–14 through A1–32 with Chapter 4, Tables A1–33 through A1–46 with Chapter 5, and Tables A1–47 through A1–50 with Chapter 6.

TABLE A1-1

PROPORTION OF WHITE MALE CANCER MORTALITY BY TYPE, UNITED STATES, 1950-1975

	(1) 1950-1954		(2) 1955-1959		(3) 1960-1964		(4) 1965-1969		(5) 1970-1975		(6) (1) - (5)
Type	%	Rank	%	Rank	%	Rank	%	Rank	%	Rank	Change in Rank
1. Lung	15.8	1	19.9	1	23.6	1	27.3	1	31.1	1	0
2. Stomach	12.5	2	9.8	3	7.7	4	5.8	4	4.6	5	-3
3. Prostate	10.4	3	10.2	2	9.7	2	9.3	3	9.3	3	0
4. Large intestine	9.5	4	9.4	4	9.4	3	9.4	2	9.5	2	2
5. Rectum	5.3	5	4.8	7	4.1	7	3.6	7	3.0	8	-3
6. Pancreas	5.1	6	5.5	5	5.7	5	5.7	5	5.4	4	2
7. Leukemia	5.0	7	5.2	6	5.2	6	5.0	6	4.5	6	1
8. Bladder	4.1	8	3.9	8	3.8	8	3.6	8	3.6	7	1
9. Liver	3.7	9	3.2	9	2.9	9	2.3	12	1.6	14	-5
10. Tongue, mouth	2.7	10	2.7	11	2.6	12	2.5	11	2.2	12	-2
11. Brain/central nervous	2.6	11	2.6	12	2.7	11	2.7	10	2.6	10	1
12. Esophagus	2.6	12	2.4	13	2.3	13	2.2	14	2.2	13	-1
13. Lymphoma	2.5	13	2.8	10	2.9	10	3.1	9	3.0	9	4
14. Kidney	2.2	14	2.2	14	2.3	14	2.3	13	2.3	11	3
15. Larynx	1.5	15	1.5	15	1.5	15	1.4	15	1.4	15	0
16. Hodgkin's disease	1.5	16	1.4	16	1.3	16	1.3	16	1.0	18	-2
17. Other skin	1.1	17	0.9	18	0.8	19	0.7	19	0.5	20	-3
18. Bone	1.1	18	0.9	19	0.7	20	0.6	20	0.6	19	-1
19. Multiple myeloma	0.8	19	1.0	17	1.1	17	1.2	17	1.3	16	3
20. Melanoma	0.8	20	0.8	20	1.0	18	1.0	18	1.1	17	3
21. Testis	0.6	21	0.5	21	0.5	21	0.5	21	0.4	22	-1
22. Lips	0.3	22	0.2	26	0.1	29	0.1	28	0.1	28	-6
23. Connective tissue	0.3	23	0.4	22	0.4	22	0.4	22	0.4	21	2
24. Nose	0.3	24	0.3	23	0.2	24	0.2	23	0.2	23	1
25. Thyroid	0.3	25	0.3	24	0.2	25	0.2	24	0.2	24	1
26. Salivary glands	0.3	26	0.3	25	0.3	23	0.2	25	0.2	25	1
27. Nasopharynx	0.2	27	0.2	27	0.2	26	0.2	26	0.2	26	1
28. Other endocrine	0.2	28	0.2	28	0.2	27	0.2	27	0.1	29	-1
29. Breast	0.2	29	0.2	29	0.2	28	0.1	29	0.2	27	2
30. Eye	0.2	30	0.1	30	0.1	30	0.1	30	0.1	30	0
31. All other types	6.5	--	6.3	--	6.3	--	6.8	--	7.2	--	--
32. Total (sum 1-31)a	100.2	--	100.1	--	100.0	--	100.0	--	100.1	--	--

aTotal may not equal 100.0 due to rounding.

TABLE A1-2

PROPORTION OF WHITE FEMALE CANCER MORTALITY BY TYPE, UNITED STATES, 1950-1975

		(1) 1950-1954		(2) 1955-1959		(3) 1960-1964		(4) 1965-1969		(5) 1970-1975		(6)
	Type	%	Rank	%	Rank	%	Rank	%	Rank	%	Rank	Change In Rank (1) - (5)
1.	Breast	18.5	1	19.2	1	19.6	1	20.0	1	20.0	1	0
2.	Large intestine	12.3	2	12.6	2	13.1	2	13.0	2	13.1	2	0
3.	Stomach	8.0	3	6.7	3	5.6	5	4.6	6	3.7	7	-4
4.	Cervix uteri	6.9	4	6.4	5	5.8	4	4.6	7	3.4	8	-4
5.	Corpus uteri	6.1	5	5.0	6	4.4	8	3.8	9	3.4	9	-4
6.	Ovary	5.9	6	6.5	4	6.6	3	6.7	3	6.6	4	2
7.	Liver	5.1	7	4.6	7	4.0	10	3.3	10	2.3	13	-6
8.	Rectum	4.2	8	4.1	10	3.7	11	3.3	11	2.9	11	-3
9.	Leukemia	3.8	9	4.3	9	4.4	9	4.4	8	4.1	6	3
10.	Pancreas	3.8	10	4.4	8	4.8	6	5.1	5	5.2	5	5
11.	Lung	3.4	11	3.9	11	4.8	7	6.7	4	9.9	3	8
12.	Bladder	2.0	12	2.0	13	1.9	14	1.8	13	1.8	14	-2
13.	Lymphoma	1.9	13	2.3	12	2.6	12	3.0	12	3.1	10	3
14.	Brain/central nervous	1.9	14	2.0	14	2.2	13	2.3	14	2.4	12	2
15.	Kidney	1.4	15	1.5	15	1.6	15	1.6	15	1.6	15	0
16.	Hodgkin's disease	1.0	16	1.0	16	1.0	17	1.0	17	0.8	20	-4
17.	Bone	0.8	17	0.7	21	0.6	21	0.5	21	0.5	21	-4
18.	Esophagus	0.8	18	0.8	18	0.8	20	0.9	20	0.9	19	-1
19.	Other skin	0.7	19	0.6	22	0.5	22	0.4	23	0.4	23	-4
20.	Tongue, mouth	0.7	20	0.8	19	0.9	18	1.0	18	1.1	17	3
21.	Melanoma	0.7	21	0.8	20	0.9	19	1.0	19	1.0	18	3
22.	Multiple myeloma	0.6	22	0.9	17	1.1	16	1.2	16	1.4	16	6
23.	Thyroid	0.6	23	0.6	23	0.5	23	0.4	22	0.4	24	-1
24.	Connective tissue	0.3	24	0.3	24	0.4	24	0.4	24	0.5	22	2
25.	Nose	0.2	25	0.2	25	0.2	25	0.2	25	0.2	26	-1
26.	Salivary glands	0.2	26	0.2	26	0.2	26	0.2	26	0.2	27	-1
27.	Larynx	0.2	27	0.2	27	0.2	27	0.2	27	0.3	25	2
28.	Eye	0.2	28	0.2	28	0.1	28	0.1	28	0.1	28	0
29.	Endocrine	0.1	29	0.1	29	0.1	29	0.1	29	0.1	29	0
30.	Nasopharynx	0.1	30	0.1	30	0.1	30	0.1	30	0.1	30	0
31.	Lips	<0.05	31	<0.05	31	<0.05	31	<0.05	31	<0.05	31	0
32.	All other types	7.6	--	7.4	--	7.5	--	8.2	--	8.7	--	--
33.	Total (sum 1-32)a	100.1	--	100.4	--	100.2	--	100.2	--	100.2	--	--

aTotal may not equal 100.0 due to rounding.

201

TABLE A1-3

PROPORTION OF NONWHITE MALE CANCER MORTALITY BY TYPE, UNITED STATES, 1950-1975

	(1) 1950-1954		(2) 1955-1959		(3) 1960-1964		(4) 1965-1969		(5) 1970-1975		(6) (1) - (5)
Type	%	Rank	%	Rank	%	Rank	%	Rank	%	Rank	Change in Rank
1. Stomach	18.2	1	14.5	2	11.7	3	9.3	3	7.1	3	-2
2. Lung	13.5	2	17.2	1	21.4	1	25.4	1	29.5	1	1
3. Prostate	13.3	3	13.9	3	13.8	2	13.2	2	13.5	2	1
4. Large intestine	6.3	4	6.1	4	6.4	4	6.5	4	6.6	4	0
5. Esophagus	4.9	5	5.1	6	5.4	6	5.5	6	5.4	6	-1
6. Liver	4.9	6	4.2	7	3.6	7	3.0	8	1.9	11	-5
7. Pancreas	4.7	7	5.6	5	5.5	5	5.7	5	5.1	5	2
8. Rectum	3.9	8	3.2	9	2.9	9	2.5	10	2.2	9	-1
9. Leukemia	3.6	9	3.7	8	3.5	8	3.3	7	3.1	7	2
10. Bladder	3.0	10	2.7	10	2.6	11	2.5	11	2.2	10	0
11. Tongue, mouth	2.5	11	2.6	11	2.7	10	2.8	9	2.6	8	3
12. Lymphoma	1.9	12	2.1	12	2.0	12	1.9	12	1.8	13	-1
13. Brain/central nervous	1.5	13	1.7	13	1.6	13	1.5	15	1.4	15	-2
14. Kidney	1.5	14	1.5	14	1.6	14	1.5	16	1.4	16	-2
15. Larynx	1.3	15	1.4	15	1.5	16	1.6	13	1.7	14	1
16. Hodgkin's disease	1.3	16	1.1	17	1.1	17	0.9	17	0.7	17	-1
17. Multiple myeloma	1.0	17	1.4	16	1.6	15	1.6	14	1.9	12	5
18. Bone	1.0	18	0.9	18	0.7	18	0.6	18	0.5	18	0
19. Other skin	0.6	19	0.5	19	0.5	19	0.4	19	0.2	22	-3
20. Nose	0.5	20	0.4	20	0.3	22	0.3	21	0.3	20	0
21. Nasopharynx	0.4	21	0.3	21	0.4	20	0.3	22	0.3	21	0
22. Connective tissue	0.3	22	0.3	22	0.4	21	0.4	20	0.4	19	3
23. Breast	0.3	23	0.2	23	0.2	23	0.2	23	0.2	23	0
24. Thyroid	0.2	24	0.2	24	0.2	24	0.2	24	0.1	25	-1
25. Testis	0.2	25	0.2	25	0.2	25	0.1	27	0.1	26	-1
26. Endocrine	0.2	26	0.2	26	0.2	26	0.1	28	0.1	27	-1
27. Salivary glands	0.2	27	0.2	27	0.2	27	0.2	25	0.1	28	-1
28. Melanoma	0.2	28	0.2	28	0.2	28	0.2	26	0.2	24	4
29. Eye	0.1	29	0.1	29	0.1	29	<0.05	29	0.1	29	0
30. Lips	0.1	30	<0.05	30	<0.05	30	<0.05	30	<0.05	30	0
31. All other types [a]	8.4	--	8.1	--	7.7	--	8.4	--	9.2	--	--
32. Total (sum 1-31)	100.0	00	99.8	--	100.2	--	100.2	--	99.9	--	--

[a] Total may not equal 100.0 due to rounding.

TABLE A1-4

PROPORTION OF NONWHITE FEMALE CANCER MORTALITY BY TYPE, UNITED STATES, 1950-1975

	Type	(1) 1950-1954 %	Rank	(2) 1955-1959 %	Rank	(3) 1960-1964 %	Rank	(4) 1965-1969 %	Rank	(5) 1970-1975 %	Rank	(6) (1) - (5) Change In Rank
1.	Cervix uteri	16.4	1	16.0	1	13.9	2	11.7	2	8.8	4	-3
2.	Breast	15.1	2	15.7	2	16.5	1	17.1	1	17.2	1	1
3.	Corpus uteri	11.8	3	8.7	3	6.9	4	5.9	5	5.0	7	-4
4.	Stomach	8.9	4	7.8	5	6.9	5	5.9	6	5.2	6	-2
5.	Large intestine	7.3	5	8.1	4	9.1	3	9.8	3	10.6	2	3
6.	Ovary	4.0	6	4.7	6	4.8	6	5.0	7	4.8	8	-2
7.	Liver	3.8	7	3.3	9	3.2	9	2.6	11	1.8	14	-7
8.	Rectum	3.6	8	3.2	10	3.1	11	2.8	10	2.5	10	-2
9.	Pancreas	2.9	9	3.9	7	4.7	7	4.9	8	5.5	5	4
10.	Lung	2.9	10	3.6	8	4.4	8	6.2	4	9.1	3	7
11.	Leukemia	2.6	11	3.0	11	3.2	10	3.2	9	3.2	9	2
12.	Bladder	2.0	12	2.0	12	2.2	12	2.1	12	2.0	13	-1
13.	Esophagus	1.2	13	1.4	13	1.7	13	1.9	13	2.3	11	2
14.	Brain/central nervous	1.1	14	1.2	15	1.5	14	1.4	16	1.5	16	-2
15.	Lymphoma	1.1	15	1.4	14	1.5	15	1.6	15	1.7	15	0
16.	Kidney	1.0	16	1.0	17	1.1	17	1.1	17	1.1	17	-1
17.	Tongue, mouth	0.9	17	0.9	18	1.0	18	1.1	18	1.1	18	-1
18.	Bone	0.9	18	0.7	19	0.6	19	0.5	20	0.5	19	-1
19.	Hodgkin's disease	0.7	19	0.6	20	0.6	20	0.6	19	0.5	20	-1
20.	Multiple myeloma	0.7	20	1.2	16	1.4	16	1.7	14	2.2	12	8
21.	Thyroid	0.5	21	0.5	21	0.5	21	0.5	21	0.4	22	-1
22.	Other skin	0.5	22	0.4	22	0.5	22	0.4	23	0.3	24	-2
23.	Connective tissue	0.2	23	0.4	23	0.5	23	0.5	22	0.5	21	2
24.	Larnynx	0.2	24	0.2	24	0.3	24	0.3	24	0.4	23	1
25.	Melanoma	0.2	25	0.2	25	0.2	25	0.3	25	0.2	25	0
26.	Salivary glands	0.2	26	0.1	27	0.1	27	0.1	27	0.1	27	-1
27.	Endocrine	0.1	27	0.1	28	0.1	28	0.1	28	0.1	28	-1
28.	Nose	0.1	28	0.2	26	0.2	26	0.2	26	0.1	29	-1
29.	Eye	0.1	29	0.1	29	0.1	29	0.1	29	0.1	30	-1
30.	Nasopharynx	0.1	30	0.1	30	0.1	30	0.1	30	0.2	26	4
31.	Lips	< 0.05	31	< 0.05	31	< 0.05	31	< 0.05	31	< 0.05	31	0
32.	All other types	9.1	--	9.1	--	9.5	--	10.2	--	11.1	--	--
33.	Total (sum 1-32)[a]	100.2	--	99.9	--	100.4	--	99.9	--	100.1	--	--

[a]Total may not equal 100.0 due to rounding.

TABLE A1-5

TOTAL POPULATION, AGE-ADJUSTED CANCER MORTALITY RATES
OF THE WHITE MALE POPULATION OF THE UNITED STATES, 1950-1975

	Type	(1) 1950-1954	(2) 1955-1959	(3) 1960-1964	(4) 1965 1969	(5) 1970-1975	Change $\frac{(5)}{(1)} \times 100$ [a]
1.	Lung	24.6	32.8	40.6	49.7	58.9	239
2.	Stomach	20.7	16.8	13.6	10.7	8.8	43
3.	Prostate	18.3	18.1	17.3	17.1	17.7	97
4.	Large intestine	15.7	16.1	16.5	17.2	18.1	115
5.	Rectum	8.6	8.1	7.2	6.6	5.7	66
6.	Pancreas	8.2	9.3	10.0	10.4	10.3	126
7.	Leukemia	7.9	8.7	9.1	9.1	8.6	109
8.	Bladder	6.8	6.8	6.7	6.6	6.8	100
9.	Liver	6.1	5.4	5.1	4.2	3.0	49
10.	Tongue, mouth	4.5	4.5	4.5	4.5	4.1	91
11.	Esophagus	4.2	4.1	4.0	4.0	4.1	98
12.	Lymphoma	3.9	4.6	5.1	5.6	5.7	146
13.	Brain/central nervous	3.9	4.2	4.5	4.8	4.9	126
14.	Kidney	3.4	3.7	3.9	4.2	4.4	129
15.	Larynx	2.4	2.5	2.5	2.6	2.6	108
16.	Hodgkin's disease	2.3	2.3	2.3	2.3	1.8	78
17.	Other skin	1.9	1.6	1.4	1.2	0.9	47
18.	Bone	1.7	1.5	1.2	1.1	1.0	59
19.	Multiple myeloma	1.2	1.7	1.9	2.1	2.5	208
20.	Melanoma	1.2	1.4	1.6	1.9	2.2	183
21.	Testis	0.9	0.8	0.8	0.8	0.8	91
22.	Lips	0.6	0.4	0.2	0.2	0.2	31
23.	Nose	0.5	0.5	0.4	0.4	0.3	71
24.	Connective tissue	0.5	0.6	0.7	0.8	0.8	167
25.	Thyroid	0.5	0.5	0.4	0.4	0.3	72
26.	Salivary glands	0.4	0.5	0.5	0.4	0.4	90
27.	Nasopharynx	0.3	0.4	0.4	0.4	0.4	122
28.	Other endocrine	0.3	0.3	0.3	0.3	0.2	75
29.	Breast	0.3	0.3	0.3	0.3	0.3	90
30.	Eye	0.3	0.2	0.2	0.2	0.2	67
31.	All Other	10.7	10.7	10.9	12.4	13.7	128
32.	Total (sum 1-31) [b]	162.7	169.2	174.1	182.4	189.5	116

[a] Change in rates computed from at least two significant digits.

[b] Total may not equal sum of all types due to rounding.

TOTAL POPULATION, AGE-ADJUSTED CANCER MORTALITY RATES
OF THE WHITE FEMALE POPULATION OF THE UNITED STATES, 1950-1975

Type	(1) 1950-1954	(2) 1955-1959	(3) 1960-1964	(4) 1965-1969	(5) 1970-1975	Change $\frac{(5)}{(1)} \times 100$ [a]
1. Breast	25.4	25.5	25.3	25.9	26.0	102
2. Large intestine	17.1	16.4	16.1	15.5	15.2	89
3. Stomach	11.1	8.7	6.9	5.4	4.3	39
4. Cervix uteri	9.5	8.6	7.5	6.1	4.5	47
5. Corpus uteri	8.4	6.5	5.5	4.7	4.1	49
6. Ovary	8.1	8.6	8.6	8.7	8.6	106
7. Liver	7.1	6.0	5.0	3.9	2.7	38
8. Rectum	5.9	5.3	4.5	3.9	3.3	56
9. Leukemia	5.5	5.8	5.8	5.7	5.2	95
10. Pancreas	5.3	5.7	5.9	6.2	6.2	117
11. Lung	4.7	5.1	6.1	8.6	12.9	274
12. Bladder	2.8	2.5	2.3	2.1	1.9	68
13. Lymphoma	2.6	3.0	3.4	3.8	3.9	150
14. Brain/central nervous	2.6	2.8	3.0	3.2	3.4	131
15. Kidney	1.9	2.0	2.0	2.0	2.0	105
16. Hodgkin's disease	1.3	1.3	1.3	1.3	1.1	85
17. Bone	1.1	0.9	0.8	0.7	0.6	56
18. Esophagus	1.1	1.0	1.0	1.1	1.1	100
19. Other skin	1.0	0.8	0.6	0.5	0.4	38
20. Tongue, mouth	1.0	1.0	1.1	1.2	1.3	133
21. Melanoma	0.9	1.1	1.2	1.3	1.4	154
22. Multiple myeloma	0.8	1.2	1.3	1.4	1.7	196
23. Thyroid	0.8	0.8	0.7	0.6	0.5	63
24. Connective tissue	0.4	0.4	0.5	0.6	0.6	169
25. Nose	0.3	0.2	0.2	0.2	0.2	55
26. Salivary glands	0.2	0.2	0.2	0.2	0.2	79
27. Larynx	0.2	0.2	0.2	0.3	0.3	139
28. Eye	0.2	0.2	0.2	0.2	0.1	64
29. Other endocrine	0.2	0.2	0.2	0.2	0.2	90
30. Nasopharynx	0.1	0.1	0.1	0.1	0.1	127
31. Lips	0.1	<0.05	<0.05	<0.05	<0.05	36
32. All other	10.5	9.7	9.3	9.9	10.4	99
33. Total (sum 1-32)[b]	138.1	132.0	126.9	125.5	124.5	90

[a]Change in rates computed from at least two significant digits.

[b]Total may not equal sum of all types due to rounding.

TOTAL POPULATION, AGE-ADJUSTED CANCER MORTALITY RATES
OF THE NONWHITE MALE POPULATION OF THE UNITED STATES, 1950-1975

	Type	(1) 1950-1954	(2) 1955-1959	(3) 1960-1964	(4) 1965-1969	(5) 1970-1975	Change $\frac{(5)}{(1)} \times 100$ [a]
1.	Stomach	28.5	25.5	22.9	19.8	16.6	58
2.	Prostate	23.0	26.5	28.8	29.6	33.0	143
3.	Lung	19.5	28.4	39.9	52.4	67.6	347
4.	Large intestine	9.9	10.7	12.5	14.0	15.6	158
5.	Liver	7.3	7.0	6.8	6.3	4.4	60
6.	Esophagus	7.3	8.5	10.0	11.2	12.3	168
7.	Pancreas	7.1	9.7	10.6	12.0	12.0	169
8.	Rectum	6.1	5.7	5.5	5.3	5.2	85
9.	Bladder	4.8	4.8	5.1	5.4	5.3	110
10.	Leukemia	4.7	5.5	6.0	6.3	6.6	140
11.	Tongue, mouth	3.8	4.3	5.0	5.6	5.9	155
12.	Lymphoma	2.5	3.3	3.5	3.8	3.9	156
13.	Kidney	2.0	2.5	2.8	3.1	3.2	160
14.	Larynx	2.0	2.3	2.8	3.2	3.8	190
15.	Brain/central nervous	1.8	2.3	2.5	2.6	2.8	156
16.	Hodgkin's disease	1.5	1.7	1.8	1.6	1.4	93
17.	Multiple myeloma	1.5	2.4	3.1	3.4	4.5	300
18.	Bone	1.4	1.4	1.2	1.1	1.0	71
19.	Other skin	0.9	0.9	0.9	0.7	0.5	58
20.	Nose	0.7	0.6	0.5	0.5	0.6	79
21.	Nasopharynx	0.5	0.5	0.6	0.6	0.6	140
22.	Breast	0.4	0.4	0.4	0.4	0.4	88
23.	Connective tissue	0.4	0.5	0.6	0.8	0.8	195
24.	Thyroid	0.4	0.3	0.4	0.3	0.3	86
25.	Testis	0.3	0.3	0.3	0.2	0.3	86
26.	Salivary glands	0.3	0.3	0.3	0.4	0.3	107
27.	Melanoma	0.2	0.4	0.4	0.5	0.5	200
28.	Other endocrine	0.2	0.2	0.2	0.2	0.2	100
29.	Eye	0.1	0.1	0.1	0.1	0.1	85
30.	Lips	0.1	0.1	0.1	< 0.05	< 0.05	33
31.	All other	12.7	13.9	14.6	17.5	21.1	166
32.	Total (sum 1-31)[b]	151.9	170.8	190.3	208.9	231.0	152

[a] Change in rates computed from at least two significant digits.

[b] Total may not equal sum of all types due to rounding.

TOTAL POPULATION, AGE-ADJUSTED CANCER MORTALITY RATES
OF THE NONWHITE FEMALE POPULATION OF THE UNITED STATES, 1950-1975

	Type	(1) 1950-1954	(2) 1955-1959	(3) 1960-1964	(4) 1965-1969	(5) 1970-1975	Change $\frac{(5)}{(1)} \times 100$ [a]
1.	Cervix uteri	21.6	21.3	18.8	15.8	12.2	56
2.	Breast	20.9	21.5	22.7	23.4	24.1	115
3.	Corpus uteri	16.9	12.5	9.9	8.3	7.0	41
4.	Stomach	13.7	11.6	10.1	8.4	7.2	53
5.	Large intestine	11.1	11.9	13.2	13.8	15.0	135
6.	Liver	5.7	4.9	4.5	3.6	2.5	44
7.	Ovary	5.5	6.4	6.6	6.8	6.7	122
8.	Rectum	5.3	4.7	4.4	3.9	3.5	66
9.	Pancreas	4.5	5.7	6.8	7.0	7.8	173
10.	Lung	4.1	5.1	6.1	8.5	12.8	312
11.	Leukemia	3.2	3.7	4.1	4.1	4.2	131
12.	Bladder	3.1	3.0	3.1	3.0	2.8	90
13.	Esophagus	1.7	1.9	2.3	2.6	3.2	188
14.	Lymphoma	1.4	1.8	2.0	2.2	2.3	164
15.	Kidney	1.3	1.3	1.5	1.5	1.5	115
16.	Brain/central nervous	1.3	1.3	1.7	1.7	1.9	146
17.	Tongue, mouth	1.3	1.2	1.4	1.5	1.6	123
18.	Bone	1.2	1.0	0.8	0.7	0.7	55
19.	Multiple myeloma	1.0	1.6	2.0	2.4	3.1	312
20.	Hodgkin's disease	0.8	0.7	0.8	0.7	0.6	70
21.	Thyroid	0.7	0.7	0.7	0.7	0.5	73
22.	Other skin	0.7	0.7	0.6	0.5	0.4	55
23.	Larynx	0.3	0.3	0.4	0.5	0.5	186
24.	Connective tissue	0.3	0.5	0.6	0.6	0.7	277
25.	Melanoma	0.2	0.3	0.3	0.3	0.3	132
26.	Salivary glands	0.2	0.2	0.2	0.2	0.2	85
27.	Nose	0.2	0.2	0.3	0.2	0.2	95
28.	Other endocrine	0.2	0.1	0.1	0.2	0.2	94
29.	Eye	0.1	0.1	0.1	0.1	0.1	80
30.	Nasopharynx	0.1	0.1	0.1	0.2	0.2	222
31.	Lips	0.1	<0.05	<0.05	<0.05	<0.05	29
32.	All other	13.1	13.0	13.5	14.2	15.5	118
33.	Total (sum 1-32) [b]	141.5	139.7	139.9	137.5	139.6	99

[a] Change in rates computed from at least two significant digits.

[b] Total may not equal sum of all types due to rounding.

TABLE A1-9

AGE-SPECIFIC ALL TYPES CANCER MORTALITY RATES UNITED STATES,

1950-1975

Mortality Rate/100,000

WHITE MALE

Age group	(1) 1950-54	(2) 1955-59	(3) 1960-64	(4) 1965-69	(5) 1970-75	(6) (5) ÷ (1) × 100	(7) (6) ÷ 116 × 100
0-4	12.4	11.8	11.6	9.3	6.9	56	48
5-9	8.8	9.5	9.1	8.8	7.2	82	71
10-14	6.4	7.0	6.9	6.6	5.5	86	74
15-19	8.9	9.0	9.5	9.7	8.0	90	78
20-24	11.2	11.3	11.3	11.3	11.0	98	84
25-29	15.0	14.9	14.5	13.7	13.1	87	75
30-34	20.9	21.5	20.8	19.6	18.6	89	77
35-39	32.0	32.7	34.2	35.6	32.0	100	86
40-44	59.2	59.9	62.5	66.4	62.0	105	91
45-49	113.1	114.5	116.5	119.8	124.1	110	95
50-54	197.2	209.1	220.5	226.7	222.6	113	97
55-59	338.7	343.9	364.3	383.9	385.7	114	98
60-64	522.3	557.8	560.5	590.6	614.1	118	102
65-69	728.2	777.5	800.9	835.1	891.1	122	105
70-74	962.6	1010.4	1066.7	1146.1	1188.3	123	106
75-84	1376.0	1406.1	1447.9	1553.8	1656.6	120	103
85+	1839.0	1872.1	1791.5	1732.5	2034.2	111	96

WHITE FEMALE

Age group	(1)	(2)	(3)	(4)	(5)	(6)	(7) (6) ÷ 90 × 100
0-4	10.6	10.1	9.5	8.4	5.6	53	59
5-9	6.9	7.2	7.2	6.8	5.4	78	87
10-14	5.2	5.3	5.5	5.3	4.3	83	92
15-19	6.2	5.9	6.5	5.9	5.3	85	94
20-24	8.0	7.3	7.2	7.1	6.3	79	88
25-29	14.5	13.6	12.1	10.8	10.7	74	82
30-34	27.8	27.6	24.0	22.0	20.5	74	82
35-39	52.1	50.0	47.8	44.9	40.3	77	86
40-44	95.0	89.0	86.6	87.2	76.7	81	90
45-49	151.0	145.7	141.8	141.2	138.1	91	101
50-54	217.9	209.5	213.4	215.2	209.4	96	107
55-59	308.2	289.1	281.6	293.1	301.0	98	109
60-64	415.3	400.7	371.4	369.7	392.7	95	106
65-69	529.1	519.0	496.0	484.7	498.9	94	104
70-74	712.6	663.7	646.9	644.5	629.6	88	98
75-84	1008.7	955.5	912.6	896.2	885.9	88	98
85+	1345.8	1315.5	1223.9	1132.9	1165.3	87	97

Age group	(1) 1950-54	(2) 1955-59	(3) 1960-64	NONWHITE MALE (4) 1965-69	(5) 1970-75	(6) (5) ÷ (1) x 100	(7) (6) ÷ 152 x 100
0-4	8.1	7.5	8.6	6.6	5.2	64	42
5-9	5.1	6.1	5.6	5.5	5.4	106	70
10-14	4.9	5.3	5.7	5.4	4.7	96	63
15-19	8.5	7.5	8.1	8.1	7.2	85	56
20-24	10.3	10.8	9.6	7.9	9.1	88	58
25-29	13.7	13.1	13.5	12.8	11.5	84	55
30-34	20.9	24.9	23.0	21.7	20.1	96	63
35-39	38.5	40.4	48.5	47.6	45.5	118	78
40-44	80.4	89.3	102.4	112.3	102.7	128	84
45-49	147.5	165.5	176.5	202.0	213.4	145	95
50-54	295.0	294.7	335.6	357.3	385.0	131	86
55-59	435.8	455.3	451.4	540.7	585.9	134	88
60-64	593.6	675.8	730.9	748.4	832.8	140	92
65-69	630.1	780.0	889.9	969.1	1014.1	161	106
70-74	787.0	897.5	1073.4	1201.6	1394.8	177	116
75-84	859.8	1005.7	1159.9	1352.2	1606.4	187	123
85+	922.7	1066.7	1148.3	1163.3	1617.2	175	115

Age group	(1) 1950-54	(2) 1955-59	(3) 1960-64	NONWHITE FEMALE (4) 1965-69	(5) 1970-75	(6) (5) ÷ (1) x 100	(7) (6) ÷ 99 x 100
0-4	6.8	6.4	7.7	6.0	4.7	69	70
5-9	4.4	4.5	5.3	4.7	4.5	102	103
10-14	3.9	4.6	5.4	4.3	4.1	105	106
15-19	6.0	6.3	5.5	5.7	5.0	83	84
20-24	8.9	8.1	7.2	7.3	6.7	75	76
25-29	18.7	18.1	16.3	14.8	13.3	71	72
30-34	46.4	41.2	36.3	34.3	26.3	57	58
35-39	83.2	75.8	75.4	67.5	57.1	69	70
40-44	154.5	141.1	130.2	117.1	108.0	70	71
45-49	212.2	204.7	191.8	191.3	184.1	87	88
50-54	345.2	308.1	299.8	272.4	272.1	79	80
55-59	399.3	405.3	351.8	357.8	361.4	91	92
60-64	489.1	470.1	520.2	463.3	463.2	95	96
65-69	444.8	489.5	511.1	561.4	530.9	119	120
70-74	553.9	571.4	608.0	615.0	735.8	133	134
75-84	605.9	633.6	678.2	719.0	784.1	129	130
85+	670.7	700.0	706.7	707.6	866.1	129	130

209

TABLE A1-10

AGE-SPECIFIC LUNG CANCER MORTALITY RATES, 1950-1954 AND 1970-1975

Mortality Rate/100,000

Age group	35-39	40-44	45-49	50-54	55-59	60-64	65-69	70-74	75-84	85+
Year										
WHITE MALE										
(1) 1950-1954	4.1	11.5	26.5	50.5	84.1	115.5	126.5	119.5	99.2	76.4
(2) 1970-1975	7.2	20.2	47.0	86.8	151.2	237.9	326.8	383.1	386.4	258.0
(3) (2) ÷ (1) × 100	176	176	177	172	180	206	258	321	390	338
(4) (3) ÷ 239 × 100	74	74	74	72	75	86	108	134	163	141
WHITE FEMALE										
(5) 1950-1954	1.4	2.8	4.9	7.3	11.0	16.4	20.6	28.2	34.6	31.2
(6) 1970-1975	3.4	8.9	18.3	29.3	42.5	51.1	55.0	55.7	60.0	61.6
(7) (6) ÷ (5) × 100	243	318	373	401	386	312	267	198	173	197
(8) (7) ÷ 274 × 100	89	116	136	146	141	114	97	72	63	72
NONWHITE MALE										
(9) 1950-1954	5.6	13.6	28.7	55.2	78.5	90.6	74.6	65.9	55.1	41.8
(10) 1970-1975	13.1	38.3	83.0	152.0	221.3	287.2	307.7	352.2	306.0	194.0
(11) (10) ÷ (9) × 100	234	282	289	275	282	317	412	534	555	464
(12) (11) ÷ 347 × 100	67	81	83	79	81	91	119	154	160	134
NONWHITE FEMALE										
(13) 1950-1954	1.8	3.9	6.8	10.6	13.8	12.2	16.0	18.6	15.3	16.0
(14) 1970-1975	4.6	12.3	22.9	32.8	40.3	45.0	45.8	59.4	52.0	50.5
(15) (14) ÷ (13) × 100	256	315	337	309	380	369	286	319	340	316
(16) (15) ÷ 312 × 100	82	101	108	99	122	118	92	102	109	101

TABLE A1-11

AGE-SPECIFIC LARGE INTESTINE AND RECTUM CANCER MORTALITY RATES,

1950-1954 AND 1970-1975

Age group	35-39	40-44	45-49	50-54	55-59	60-64	65-69	70-74	75-84	85+
Year										
WHITE MALE										
(1) 1950-1954	3.8	13.0	13.7	24.9	45.5	75.8	114.7	162.0	237.3	299.7
(2) 1970-1975	2.9	5.4	11.5	22.7	41.5	70.3	109.6	159.7	254.9	349.0
(3) (2) ÷ (1) × 100	76	45	84	91	91	93	96	99	107	116
(4) (3) ÷ 98 × 100	78	46	86	93	93	95	98	101	109	118
WHITE FEMALE										
(5) 1950-1954	4.7	9.5	16.8	28.7	45.6	67.9	95.2	136.8	221.6	319.6
(6) 1970-1975	2.7	5.8	11.3	20.9	35.4	53.8	79.8	116.7	189.9	281.1
(7) (6) ÷ (5) × 100	57	61	67	73	78	79	84	85	86	88
(8) (7) ÷ 80 × 100	71	76	84	91	98	99	105	106	108	110
NONWHITE MALE										
(9) 1950-1954	3.9	8.0	13.4	25.6	39.6	59.1	72.8	95.2	110.4	114.6
(10) 1970-1975	3.9	6.9	13.2	25.3	40.7	69.1	96.3	140.7	184.4	199.2
(11) (10) ÷ (9) × 100	100	86	99	99	103	117	132	148	167	174
(12) (11) ÷ 130 × 100	77	66	76	76	79	90	102	114	128	134
NONWHITE FEMALE										
(13) 1950-1954	5.4	11.5	17.9	35.0	47.2	61.9	61.0	76.7	95.7	120.0
(14) 1970-1975	3.5	7.3	14.3	25.2	42.0	60.5	81.7	123.8	152.4	177.1
(15) (14) ÷ (13) × 100	65	63	80	72	89	98	134	161	159	148
(16) (15) ÷ 113 × 100	58	56	71	64	79	87	119	142	141	131

TABLE A1-12

AGE-SPECIFIC FEMALE BREAST CANCER MORTALITY RATES,

1950-1954 AND 1970-1975

	Age group	35-39	40-44	45-49	50-54	55-59	60-64	65-69	70-74	75-84	85+
	Year										
					WHITE FEMALE						
(1)	1950-1954	13.7	27.1	42.0	53.2	66.3	77.6	86.1	108.3	142.1	202.5
(2)	1970-1975	12.5	24.6	43.5	60.1	76.0	85.2	93.8	104.4	129.6	170.8
(3)	(2) ÷ (1) × 100	91	91	104	113	115	110	109	96	91	84
(4)	(3) ÷ 102 × 100	89	89	102	111	113	108	107	94	89	82
					NONWHITE FEMALE						
(5)	1950-1954	16.6	29.7	37.5	54.3	58.4	65.4	56.9	72.3	80.7	100.0
(6)	1970-1975	16.2	29.2	45.2	62.2	68.2	74.3	77.2	93.1	93.7	118.1
(7)	(6) ÷ (5) × 100	98	98	121	115	117	114	136	129	116	118
(8)	(7) ÷ 115 × 100	85	85	105	100	102	99	118	112	101	103

TABLE A1-13

AGE-SPECIFIC BLADDER CANCER MORTALITY RATES,

1950-1954 AND 1970-1975

Age group		45-49	50-54	55-59	60-64	65-69	70-74	75-84	85+
	Year								
	WHITE MALE								
(1)	1950-1954	2.5	5.8	12.1	21.0	32.4	46.2	74.0	113.0
(2)	1970-1975	1.6	3.9	8.2	16.8	30.4	51.0	86.4	132.8
(3)	(2) ÷ (1) × 100	64	67	68	80	94	110	117	118
(4)	(3) ÷ 100 × 100	64	67	68	80	94	110	117	118
	WHITE FEMALE								
(5)	1950-1954	1.1	2.0	3.7	6.6	11.1	19.9	34.5	50.8
(6)	1970-1975	0.6	1.3	2.6	4.7	7.4	12.6	25.5	47.6
(7)	(6) ÷ (5) × 100	55	65	70	71	67	63	74	94
(8)	(7) ÷ 68 × 100	81	96	103	104	99	93	109	138
	NONWHITE MALE								
(9)	1950-1954	3.8	8.9	13.6	17.7	19.0	32.9	29.1	37.3
(10)	1970-1975	2.1	5.6	10.9	17.4	26.8	38.9	52.4	56.8
(11)	(10) ÷ (9) × 100	55	63	80	98	141	118	180	152
(12)	(11) ÷ 110 × 100	50	57	73	89	128	107	164	138
	NONWHITE FEMALE								
(13)	1950-1954	3.5	6.3	8.8	10.7	13.0	17.7	17.2	18.7
(14)	1970-1975	1.7	3.6	5.8	9.3	14.3	20.2	24.7	31.2
(15)	(14) ÷ (13) × 100	49	57	66	87	110	114	144	167
(16)	(15) ÷ 90 × 100	54	63	73	97	122	127	160	186

TABLE A1-14

COMPARISON OF AGE-ADJUSTED CANCER MORTALITY RATES IN COUNTIES AT OPPOSITE ENDS OF THE URBANIZATION SPECTRUM, 1950-1969

	≥ 98% URBAN			0% URBAN			$\frac{(1)}{(4)}\times100$	$\frac{(2)}{(5)}\times100$	$\frac{(3)}{(6)}\times100$
	(1)	(2)	(3)	(4)	(5)	(6)	(7)	(8)	(9)
	1950-54	1965-69	1950-69	1950-54	1965-69	1950-69			
WHITE MALE									
1. All types	199.1	205.4	201.7	130.9	154.9	141.5	152	133	143
2. Digestive	80.8	63.9	71.6	50.5	44.3	47.2	160	144	152
(Large intestine)	(20.1)	(21.1)	(20.6)	(10.9)	(13.6)	(12.3)	(184)	(155)	(167)
3. Respiratory	38.1	59.5	49.4	17.7	37.5	27.7	215	159	178
4. Reproductive	19.5	17.9	18.5	17.1	21.5	18.3	114	83	101
5. Urinary	13.5	12.5	12.9	7.1	8.1	7.8	190	154	165
(Bladder)	(9.2)	(7.9)	(8.5)	(4.2)	(5.0)	(4.7)	(219)	(158)	(181)
6. Lymphatic	17.2	19.8	18.9	13.4	17.2	15.6	128	115	121
7. All other	30.0	31.8	30.4	25.1	26.3	24.9	120	121	122
WHITE FEMALE									
1. All types	155.0	138.9	144.9	123.7	119.9	116.8	125	116	124
2. Digestive	55.3	40.4	47.1	43.6	37.6	38.2	127	107	123
(Large intestine)	(19.6)	(17.0)	(18.3)	(13.5)	(15.4)	(14.7)	(145)	(110)	(124)
3. Respiratory	6.3	10.9	8.0	4.3	7.5	5.5	147	145	145
4. Breast	30.4	30.1	30.0	17.9	21.5	18.7	170	140	160
5. Reproductive	26.3	20.1	22.9	24.5	18.7	21.0	107	107	109
6. Urinary	5.3	4.5	4.9	3.7	4.6	4.5	143	98	109
(Bladder)	(3.1)	(2.3)	(2.7)	(1.6)	(2.0)	(2.1)	(194)	(115)	(129)
7. Lymphatic	11.5	12.9	12.6	9.7	11.9	10.4	119	108	121
8. All other	19.9	20.0	19.4	20.0	18.1	18.5	100	110	105

214

PROPORTION OF WHITE MALE CANCER MORTALITY BY CANCER TYPE, STRONGLY URBAN
COUNTIES OF THE UNITED STATES, 1950-1975

Type	1950-1954		1960-1964		1970-1975		Change in Rank 1950-1954 to 1970-1975
	%	Rank	%	Rank	%	Rank	
1. Lung	17.6	1	24.8	1	31.1	1	0
2. Stomach	11.8	2	7.6	4	4.8	5	-3
3. Large intestine	9.7	3	9.6	2	9.9	2	1
4. Prostate	8.8	4	8.4	3	8.6	3	1
5. Rectum	5.8	5	4.4	7	3.1	8	-3
6. Pancreas	5.0	6	5.6	5	5.3	4	2
7. Leukemia	4.6	7	4.9	6	4.3	6	1
8. Bladder	4.3	8	3.9	8	3.7	7	1
9. Liver	3.4	9	2.9	9	1.6	14	-5
10. Tongue, mouth	3.1	10	2.9	10	2.4	11	-1
11. Esophagus	3.0	11	2.6	13	2.3	13	-2
12. Brain/central nervous	2.6	12	2.7	12	2.6	10	2
13. Lymphoma	2.5	13	2.9	11	3.0	9	4
14. Kidney	2.2	14	2.3	14	2.3	12	2
15. Larynx	1.8	15	1.6	15	1.5	15	0
16. Hodgkin's disease	1.5	16	1.3	16	1.0	18	-2
17. Multiple myeloma	0.8	17	1.0	17	1.2	16	1
18. Melanoma	0.8	18	0.9	18	1.2	17	1

PROPORTION OF WHITE MALE CANCER MORTALITY BY CANCER TYPE, MODERATELY URBAN
COUNTIES OF THE UNITED STATES, 1950-1975

Type	1950-1954 %	Rank	1960-1964 %	Rank	1970-1975 %	Rank	Change In Rank 1950-1954 to 1970-1975
1. Lung	14.4	1	22.9	1	31.5	1	0
2. Stomach	13.0	2	7.6	4	4.4	6	-4
3. Prostate	11.6	3	10.6	2	9.7	2	1
4. Large intestine	9.4	4	9.2	3	9.3	3	1
5. Leukemia	5.3	5	5.5	6	4.7	5	0
6. Pancreas	5.2	6	5.7	5	5.4	4	2
7. Rectum	4.8	7	4.0	7	3.0	8	-1
8. Bladder	3.9	8	3.7	8	3.5	7	1
9. Liver	3.9	9	2.9	10	1.5	14	-5
10. Brain/central nervous	2.5	10	2.6	11	2.6	10	0
11. Lymphoma	2.5	11	3.0	9	3.0	9	2
12. Tongue, mouth	2.4	12	2.3	13	2.0	13	-1
13. Esophagus	2.3	13	2.1	14	2.0	12	1
14. Kidney	2.1	14	2.4	12	2.3	11	3
15. Hodgkin's disease	1.5	15	1.4	15	1.0	18	-3
16. Larynx	1.3	16	1.3	16	1.3	15	1
17. Other skin	1.2	17	0.9	18	0.5	20	-3
18. Bone	1.2	18	0.7	20	0.6	19	-1
19. Melanoma	0.8	19	0.9	19	1.1	17	2
20. Multiple myeloma	0.8	20	1.1	17	1.3	16	4

PROPORTION OF WHITE MALE CANCER MORTALITY BY CANCER TYPE, RURAL COUNTIES OF THE
UNITED STATES, 1970-1975

Type		1950-1954		1960-1964		1970-1975		Change in Rank 1950-1954 to 1970-1975
		%	Rank	%	Rank	%	Rank	
1.	Stomach	13.9	1	8.1	4	4.4	6	-5
2.	Prostate	13.3	2	12.1	2	10.6	2	0
3.	Lung	12.6	3	21.0	1	30.9	1	2
4.	Large intestine	9.0	4	8.9	3	8.9	3	1
5.	Leukemia	5.5	5	5.9	6	4.9	5	0
6.	Pancreas	5.1	6	6.2	5	5.6	4	2
7.	Rectum	4.4	7	3.6	7	2.8	9	-2
8.	Liver	4.3	8	3.1	9	1.5	14	-6
9.	Bladder	3.6	9	3.6	8	3.4	7	2
10.	Brain/central nervous	2.6	10	2.6	11	2.6	10	0
11.	Lymphoma	2.4	11	3.0	10	3.0	8	3
12.	Tongue,mouth	2.2	12	2.0	13	1.8	13	-1
13.	Kidney	2.1	13	2.3	12	2.3	11	2
14.	Esophagus	1.7	14	1.8	14	1.9	12	2
15.	Other skin	1.6	15	1.2	17	0.6	19	-4
16.	Hodgkin's disease	1.5	16	1.4	15	0.9	18	-2
17.	Bone	1.3	17	0.8	20	0.6	20	-3
18.	Larynx	1.0	18	1.2	18	1.2	16	2
19.	Melanoma	0.8	19	1.0	19	1.1	17	2
20.	Multiple myeloma	0.7	20	1.2	16	1.4	15	5

PROPORTION OF WHITE FEMALE CANCER MORTALITY BY CANCER TYPE, STRONGLY URBAN
COUNTIES OF THE UNITED STATES, 1950-1975

Type	1950-1954		1960-1964		1970-1975		Change In Rank 1950-1954 to 1970-1975
	%	Rank	%	Rank	%	Rank	
1. Breast	19.5	1	20.5	1	20.7	1	0
2. Large intestine	12.4	2	13.2	2	12.8	2	0
3. Stomach	7.8	3	5.7	4	3.8	7	-4
4. Cervix uteri	6.6	4	5.2	5	3.0	10	-6
5. Ovary	6.3	5	6.8	3	6.6	4	1
6. Corpus uteri	5.5	6	4.1	9	3.3	8	-2
7. Liver	4.6	7	3.8	10	2.2	13	-6
8. Rectum	4.6	8	3.8	11	2.9	11	-3
9. Leukemia	3.9	9	4.2	8	3.9	6	3
10. Pancreas	3.8	10	4.7	7	5.2	5	5
11. Lung	3.5	11	5.1	6	10.7	3	8
12. Bladder	2.0	12	1.9	14	1.8	14	-2
13. Lymphoma	2.0	13	2.7	12	3.1	9	4
14. Brain/central nervous	1.9	14	2.2	13	2.4	12	2
15. Kidney	1.3	15	1.5	15	1.5	15	0
16. Hodgkin's disease	1.0	16	1.0	17	0.8	18	-2
17. Tongue ,mouth	0.7	17	0.9	18	1.1	17	0
18. Multiple myeloma	0.6	18	1.1	16	1.3	16	2

TABLE A1-19

PROPORTION OF WHITE FEMALE CANCER MORTALITY BY CANCER TYPE, MODERATELY URBAN COUNTIES OF THE UNITED STATES, 1950-1975

Type	1950-1954 %	Rank	1960-1964 %	Rank	1970-1975 %	Rank	Change In Rank 1950-1954 to 1970-1975
1. Breast	17.6	1	18.7	1	19.5	1	0
2. Large intestine	12.0	2	13.0	2	13.4	2	0
3. Stomach	7.8	3	5.4	5	3.5	9	-6
4. Cervix uteri	7.8	4	6.4	4	3.8	7	-3
5. Corpus uteri	6.7	5	4.7	7	3.5	8	-3
6. Ovary	5.7	6	6.6	3	6.7	4	2
7. Liver	5.5	7	4.1	10	2.4	12	-5
8. Leukemia	4.0	8	4.5	8	4.2	6	2
9. Rectum	3.9	9	3.6	11	2.9	11	-2
10. Pancreas	3.8	10	4.8	6	5.3	5	5
11. Lung	3.2	11	4.3	9	9.3	3	8
12. Bladder	2.0	12	2.0	14	1.8	14	-2
13. Lymphoma	1.8	13	2.6	12	3.1	10	3
14. Brain/central nervous	1.8	14	2.2	13	2.4	13	1
15. Kidney	1.5	15	1.6	15	1.6	15	0
16. Hodgkin's disease	1.0	16	1.0	17	0.9	18	-2
17. Melanoma	0.7	17	0.9	18	1.1	17	0
18. Multiple myeloma	0.6	18	1.1	16	1.5	16	2

PROPORTION OF WHITE FEMALE CANCER MORTALITY BY CANCER TYPE, RURAL COUNTIES OF
THE UNITED STATES, 1950-1975

	Type	1950-1954 %	Rank	1960-1964 %	Rank	1970-1975 %	Rank	Change In Rank 1950-1954 to 1970-1975
1.	Breast	16.7	1	17.9	1	18.6	1	0
2.	Large intestine	12.3	2	12.9	2	13.4	2	0
3.	Stomach	8.5	3	5.7	5	3.7	8	-5
4.	Cervix uteri	7.1	4	6.5	3	4.0	7	-3
5.	Corpus uteri	7.1	5	4.7	8	3.6	9	-4
6.	Liver	6.9	6	4.6	9	2.5	12	-6
7.	Ovary	5.2	7	6.3	4	6.5	4	3
8.	Leukemia	4.0	8	4.7	7	4.5	6	2
9.	Pancreas	3.8	9	4.9	6	5.3	5	4
10.	Rectum	3.6	10	3.4	11	2.8	11	-1
11.	Lung	3.3	11	4.2	10	8.6	3	8
12.	Bladder	1.9	12	1.8	14	1.8	14	-2
13.	Brain/central nervous	1.7	13	2.0	13	2.4	13	0
14.	Lymphoma	1.6	14	2.5	12	3.1	10	4
15.	Kidney	1.4	15	1.6	15	1.7	15	0
16.	Other skin	1.1	16	0.7	19	0.5	19	-3
17.	Hodgkin's disease	0.9	17	1.0	18	0.7	18	-1
18.	Melanoma	0.7	18	1.0	17	1.1	17	1
19.	Multiple myeloma	0.6	19	1.1	16	1.5	16	3

TABLE A1-21

TOTAL POPULATION AGE-ADJUSTED CANCER MORTALITY RATES OF THE WHITE MALE POPULATION, STRONGLY URBAN COUNTIES OF THE UNITED STATES, 1950- 1975

Type	1950 - 1954				1960 - 1964				1970 - 1975			
	Rate	SU[a]/USA	SU/RU	SU/MU	Rate	SU/USA	SU/RU	SU/MU	Rate	SU/USA	SU/RU	SU/MU
1. Lung	30.0	122	176	144	45.7	113	141	121	61.1	104	111	104
2. Stomach	22.1	107	118	113	14.7	108	121	118	9.5	108	123	122
3. Prostate	18.5	101	102	103	17.2	99	98	99	17.6	99	98	99
4. Large intestine	18.2	116	149	129	18.6	113	141	123	19.8	109	128	116
5. Rectum	10.7	124	181	147	8.5	118	160	131	6.3	111	131	113
6. Pancreas	9.1	111	132	118	10.6	106	114	113	10.6	103	107	106
7. Leukemia	8.3	105	112	108	9.1	100	101	101	8.5	99	98	98
8. Bladder	8.3	122	169	138	8.5	127	160	142	7.4	109	128	116
9. Liver	6.3	103	109	109	5.4	106	115	113	3.3	110	127	122
10. Tongue, mouth	5.6	124	193	156	5.5	122	183	149	4.7	115	152	127
11. Esophagus	5.6	133	243	165	4.9	123	181	144	4.6	112	139	124
12. Lymphoma	4.3	110	130	119	5.3	104	115	108	6.0	105	113	109
13. Brain/central nervous	4.2	108	120	117	4.7	104	109	109	5.1	104	104	106
14. Kidney	3.9	115	139	126	4.2	108	120	108	4.5	102	107	105
15. Larynx	3.2	133	229	160	3.1	124	172	141	2.9	112	132	116
16. Hodgkin's disease	2.4	104	114	109	2.4	104	109	109	1.8	100	100	100
17. Other skin	1.6	84	73	84	1.3	93	76	87	0.8	89	80	89
18. Bone	1.8	106	100	106	1.2	100	100	100	1.0	100	100	91
19. Multiple myeloma	1.4	117	140	127	1.9	100	106	100	2.5	100	100	104
20. Melanoma	1.3	108	130	118	1.7	106	100	106	2.3	105	110	115
21. Total	182.7	112	135	123	189.1	109	125	116	197.9	104	112	107

USA= United States   [a] All ratios have been multiplied by 100.
SU = Strongly urban counties
MU = Moderately urban counties
RU = Rural counties

221

TABLE A1-22

TOTAL POPULATION AGE-ADJUSTED CANCER MORTALITY RATES OF THE WHITE MALE POPULATION, MODERATELY URBAN COUNTIES OF THE UNITED STATES, 1950-1975

	Type	1950 - 1954 Rate	$\frac{MU}{USA}$[a]	$\frac{MU}{RU}$	1960 - 1964 Rate	$\frac{MU}{USA}$	$\frac{MU}{RU}$	1970 - 1975 Rate	$\frac{MU}{USA}$	$\frac{MU}{RU}$
1.	Lung	20.9	85	123	37.7	93	116	58.5	99	106
2.	Stomach	19.5	94	104	12.5	92	103	8.2	93	106
3.	Prostate	18.0	98	99	17.4	101	99	17.7	100	99
4.	Large intestine	14.1	90	116	15.1	92	114	17.1	94	110
5.	Rectum	7.3	85	124	6.5	90	123	5.6	98	117
6.	Pancreas	7.7	94	112	9.4	94	101	10.0	97	101
7.	Leukemia	7.7	97	104	9.0	99	100	8.7	101	100
8.	Bladder	6.0	88	122	6.0	90	113	6.4	94	110
9.	Liver	5.8	95	100	4.8	94	102	2.7	90	104
10.	Tongue, mouth	3.6	80	124	3.7	82	123	3.7	90	119
11.	Esophagus	3.4	81	148	3.4	85	126	3.7	90	112
12.	Lymphoma	3.6	92	109	4.9	96	107	5.5	96	104
13.	Brain/central nerv.	3.6	92	103	4.3	96	100	4.8	98	98
14.	Kidney	3.1	91	111	3.9	100	111	4.3	98	102
15.	Larynx	2.0	83	143	2.2	88	122	2.5	96	114
16.	Hodgkin's disease	2.2	96	105	2.2	96	100	1.8	100	100
17.	Other skin	1.9	100	86	1.5	107	88	0.9	100	90
18.	Bone	1.7	100	94	1.2	100	100	1.1	110	110
19.	Multiple myeloma	1.1	92	110	1.9	100	106	2.4	100	96
20.	Melanoma	1.1	92	110	1.6	100	94	2.0	91	95
21.	Total	148.9	92	110	163.7	94	108	184.7	98	105

[a]All ratios have been multiplied by 100.

USA= United States
SU = Strongly urban counties
MU = Moderately urban counties
RU = Rural counties

TOTAL POPULATION AGE-ADJUSTED CANCER MORTALITY RATES OF THE WHITE MALE POPULATION, RURAL COUNTIES OF THE UNITED STATES, 1950 - 1975

	Type	1950 - 1954 Rate	1950 - 1954 $\frac{RU}{USA}$[a]	1960 - 1964 Rate	1960 - 1964 $\frac{RU}{USA}$	1970 - 1975 Rate	1970 - 1975 $\frac{RU}{USA}$
1.	Lung	17.0	69	32.4	80	55.1	94
2.	Stomach	18.8	91	12.1	89	7.7	88
3.	Prostate	18.1	99	17.5	101	17.9	101
4.	Large intestine	12.2	78	13.2	80	15.5	86
5.	Rectum	5.9	69	5.3	74	4.8	84
6.	Pancreas	6.9	84	9.3	93	9.9	96
7.	Leukemia	7.4	94	9.0	99	8.7	101
8.	Bladder	4.9	72	5.3	79	5.8	85
9.	Liver	5.8	95	4.7	92	2.6	87
10.	Tongue,mouth	2.9	64	3.0	67	3.1	76
11.	Esophagus	2.3	55	2.7	68	3.3	80
12.	Lymphoma	3.3	85	4.6	90	5.3	93
13.	Brain/central nervous	3.5	90	4.3	96	4.9	100
14.	Kidney	2.8	82	3.5	90	4.2	95
15.	Larynx	1.4	58	1.8	72	2.2	85
16.	Hodgkin's disease	2.1	91	2.2	96	1.8	100
17.	Other skin	2.2	116	1.7	121	1.1	122
18.	Bone	1.8	106	1.2	100	1.0	100
19.	Multiple myeloma	1.0	83	1.8	95	2.5	104
20.	Melanoma	1.0	83	1.7	106	2.1	95
21.	Total	135.1	83	151.5	87	176.2	93

[a]All ratios have been multiplied by 100.

USA= United States
SU = Strongly urban counties
MU = Moderately urban counties
RU = Rural counties

223

TABLE A1-24

TOTAL POPULATION AGE-ADJUSTED CANCER MORTALITY RATES OF THE WHITE FEMALE POPULATION, STRONGLY URBAN COUNTIES OF THE UNITED STATES, 1950-1975

Type	1950 - 1954				1960 - 1964				1970 - 1975			
	Rate	$\frac{SU}{USA}$[a]	$\frac{SU}{RU}$	$\frac{SU}{MU}$	Rate	$\frac{SU}{USA}$	$\frac{SU}{RU}$	$\frac{SU}{MU}$	Rate	$\frac{SU}{USA}$	$\frac{SU}{RU}$	$\frac{SU}{MU}$
1. Breast	28.1	111	133	122	27.5	109	127	119	27.9	107	122	114
2. Large intestine	18.4	108	120	117	17.2	107	119	114	15.7	103	110	105
3. Stomach	11.7	105	111	116	7.3	106	118	118	4.6	107	118	121
4. Cervix uteri	9.4	99	103	92	7.1	95	87	88	4.1	91	79	82
5. Corpus uteri	8.0	95	90	92	5.5	100	102	96	4.2	102	102	102
6. Ovary	9.0	111	136	120	9.1	106	118	110	8.9	103	113	105
7. Liver	6.9	97	93	96	5.0	100	98	104	2.8	104	100	104
8. Rectum	6.8	115	151	133	5.0	111	132	119	3.5	106	117	106
9. Leukemia	5.8	105	114	109	5.9	102	104	105	5.2	100	98	102
10. Pancreas	5.7	108	121	116	6.2	105	113	111	6.4	103	112	107
11. Lung	5.2	111	124	124	6.8	111	136	128	14.3	111	138	122
12. Bladder	3.0	107	130	115	2.4	104	120	109	2.0	105	111	105
13. Lymphoma	2.9	112	138	121	3.6	106	120	113	4.0	103	111	108
14. Brain/central nerv.	2.8	108	127	122	3.1	103	115	107	3.5	103	106	106
15. Kidney	2.0	105	111	100	2.0	100	105	100	2.0	100	100	105
16. Hodgkin's disease	1.4	108	127	117	1.4	108	117	108	1.1	100	110	100
17. Melanoma	0.9	100	100	100	1.1	92	85	92	1.4	100	100	100
18. Multiple myeloma	0.9	113	129	113	1.4	108	108	108	1.7	100	100	100
19. Total	146.2	106	116	112	133.1	105	114	110	129.7	104	112	108

[a] All ratios have been multiplied by 100.

USA= United States
SU = Strongly urban counties
MU = Moderately urban counties
RU = Rural counties

224

TOTAL POPULATION AGE-ADJUSTED CANCER MORTALITY RATES OF THE WHITE FEMALE POPULATION, MODERATELY URBAN COUNTIES OF THE UNITED STATES, 1950-1975

	Type	1950 - 1954 Rate	$\frac{MU^a}{USA}$	$\frac{MU}{RU}$	1960 - 1964 Rate	$\frac{MU}{USA}$	$\frac{MU}{RU}$	1970 - 1975 Rate	$\frac{MU}{USA}$	$\frac{MU}{RU}$
1.	Breast	23.1	91	109	23.2	92	107	24.5	94	107
2.	Large intestine	15.7	92	103	15.1	94	105	15.0	99	105
3.	Stomach	10.1	91	96	6.2	90	100	3.8	88	97
4.	Cervix uteri	10.2	107	112	8.1	108	99	5.0	111	96
5.	Corpus uteri	8.7	104	98	5.7	104	106	4.1	100	100
6.	Ovary	7.5	93	114	8.3	97	108	8.5	99	108
7.	Liver	7.2	101	97	4.8	96	94	2.7	100	96
8.	Rectum	5.1	86	113	4.2	93	111	3.3	100	110
9.	Leukemia	5.3	96	104	5.6	97	98	5.1	98	96
10.	Pancreas	4.9	92	104	5.6	95	102	6.0	97	105
11.	Lung	4.2	89	100	5.3	87	106	11.7	91	113
12.	Bladder	2.6	93	113	2.2	96	110	1.9	100	106
13.	Lymphoma	2.4	92	114	3.2	94	107	3.7	95	103
14.	Brain/central ner.	2.3	88	105	2.9	97	107	3.3	97	100
15.	Kidney	2.0	105	111	2.0	100	105	1.9	95	95
16.	Hodgkin's disease	1.2	92	109	1.3	100	108	1.1	100	110
17.	Melanoma	0.9	100	100	1.2	100	92	1.4	100	100
18.	Multiple myeloma	0.8	100	114	1.3	100	100	1.7	100	100
19.	Total	131.0	95	104	120.9	95	103	120.4	97	104

[a]All ratios have been multiplied by 100.

USA= United States
SU = Strongly urban counties
MU = Moderately urban counties
RU = Rural counties

TOTAL POPULATION AGE-ADJUSTED CANCER MORTALITY RATES OF THE WHITE FEMALE
POPULATION, RURAL COUNTIES OF THE UNITED STATES,   1950 - 1975

	Type	1950 -1954 Rate	1950 -1954 $\frac{RU}{USA}$[a]	1960 - 1964 Rate	1960 - 1964 $\frac{RU}{USA}$	1970 - 1975 Rate	1970 - 1975 $\frac{RU}{USA}$
1.	Breast	21.2	83	21.6	85	22.8	88
2.	Large intestine	15.3	89	14.4	89	14.3	94
3.	Stomach	10.5	95	6.2	90	3.9	91
4.	Cervix uteri	9.1	96	8.2	109	5.2	116
5.	Corpus uteri	8.9	106	5.4	98	4.1	100
6.	Ovary	6.6	81	7.7	89	7.9	92
7.	Liver	7.4	104	5.1	102	2.8	104
8.	Rectum	4.5	76	3.8	84	3.0	91
9.	Leukemia	5.1	93	5.7	98	5.3	102
10.	Pancreas	4.7	89	5.5	93	5.7	92
11.	Lung	4.2	89	5.0	82	10.4	81
12.	Bladder	2.3	82	2.0	87	1.8	95
13.	Lymphoma	2.1	81	3.0	88	3.6	92
14.	Brain/central nervous	2.2	85	2.7	90	3.3	97
15.	Kidney	1.8	95	1.9	95	2.0	100
16.	Hodgkin's disease	1.1	85	1.2	92	1.0	91
17.	Melanoma	0.9	100	1.3	108	1.4	100
18.	Multiple myeloma	0.7	88	1.3	100	1.7	100
19.	Total	125.6	91	117.0	92	115.5	93

[a]All ratios have been multiplied by 100.

USA= United States
SU = Strongly urban counties
MU = Moderately urban counties
RU = Rural counties

TABLE A1-27

RANK CORRELATION OF CHANGE IN CANCER MORTALITY RATES BETWEEN THE STRONGLY URBAN,
MODERATELY URBAN, RURAL COUNTIES AND THE USA, 1950-1954 TO 1970-1975,
WHITE MALES

Region	Strongly Urban Correlation[a]	Moderately Urban Correlation[a]	Rural Correlation[a]	United States Correlation[a]
Strongly Urban	---	.97	.91	.97
Moderately Urban	---	---	.97	.99
Rural	---	---	---	.94
United States	---	---	---	---

[a]Twenty-one types of cancer were included.

227

TABLE A1-28

RANK CORRELATION OF CHANGE IN CANCER MORTALITY RATES BETWEEN THE STRONGLY
URBAN, MODERATELY URBAN, RURAL COUNTIES AND THE USA, 1950 - 1954 TO 1970 - 1975

WHITE FEMALES

Region	Strongly Urban Correlation[a]	Moderately Urban Correlation[a]	Rural Correlation[a]	United States Correlation[a]
Strongly Urban	---	.97	.99	.99
Moderately Urban	---	---	.99	.98
Rural	---	---	---	.99
United States	---	---	---	---

[a]
Nineteen types were included.

TABLE A1-29

RANK CORRELATION OF CHANGE IN CANCER MORTALITY RATES, 1950 - 1954 TO 1970 - 1975

WHITE MALES AND WHITE FEMALES

Region	Correlation[a]
Strongly Urban	.99
Moderately Urban	.94
Rural	.97

[a]
Fifteen types were included.

228

TABLE A1-30

AGE-SPECIFIC CANCER MORTALITY RATES STRONGLY URBAN COUNTIES, 1950 - 1975

SU = Strongly urban
USA= United States

WHITE MALE

Year	Mortality rate/100,000																
	0-4	5-9	10-14	15-19	20-24	25-29	30-34	35-39	40-44	45-49	50-54	55-59	60-64	65-69	70-74	75-84	85+
(1) 1950-1954	12.8	9.3	6.9	9.7	11.3	15.1	21.2	33.7	63.7	124.1	220.8	386.7	602.2	852.6	1104.0	1510.3	1921.2
SU/USA x 100	103	106	108	109	101	101	101	105	108	110	112	114	115	117	115	110	105
(2) 1960-1964	11.7	9.8	7.3	9.7	10.8	14.3	21.0	34.6	63.3	122.3	234.1	393.1	615.2	887.4	1186.7	1586.3	1862.0
SU/USA x 100	101	108	106	102	96	99	101	101	101	105	106	108	110	111	111	110	104
(3) 1970-1975	6.9	7.3	5.7	8.1	11.0	13.2	18.4	31.6	62.3	124.4	224.5	394.2	633.9	935.9	1257.8	1776.0	2160.6
SU/USA x 100	100	101	104	101	100	101	99	99	100	100	101	102	103	105	106	107	106
(3) ÷ (1) x 100	54	78	83	84	97	87	99	94	98	100	102	102	105	110	114	118	112

WHITE FEMALE

Year	Mortality rate/ 100,000																
	0-4	5-9	10-14	15-19	20-24	25-29	30-34	35-39	40-44	45-49	50-54	55-59	60-64	65-69	70-74	75-84	85+
(4) 1950-1954	11.1	7.5	5.5	6.2	8.1	14.2	27.6	53.7	99.5	159.5	232.4	335.4	447.0	568.5	755.3	1057.8	1359.7
SU/USA x 100	105	109	106	100	101	98	99	103	105	106	107	109	108	107	106	105	101
(5) 1960-1964	9.7	7.9	5.7	6.5	7.2	11.9	24.0	48.2	89.3	149.2	225.2	296.7	393.8	530.9	681.7	948.0	1254.5
US/USA x 100	102	110	104	100	100	98	100	101	103	105	106	105	106	107	105	104	103
(6) 1970-1975	5.6	5.6	4.3	5.5	6.1	10.6	20.4	40.2	78.0	142.4	218.9	314.7	415.4	525.3	659.7	921.2	1192.9
SU/USA x 100	100	104	100	104	97	99	100	100	102	103	105	105	106	105	105	104	102
(6) ÷ (4) x 100	50	75	78	89	75	75	74	75	78	89	94	94	93	92	87	87	88

229

TABLE A1-31

AGE-SPECIFIC CANCER MORTALITY RATES MODERATELY URBAN COUNTIES, 1950-1975

W H I T E   M A L E

Mortality rate/100,000

Year	0-4	5-9	10-14	15-19	20-24	25-29	30-34	35-39	40-44	45-49	50-54	55-59	60-64	65-69	70-74	75-84	85+
(1) 1950-1954	11.9	8.3	5.7	7.8	10.2	13.7	20.0	30.9	56.2	104.5	174.7	304.5	461.7	649.0	884.3	1295.1	1767.7
MU/USA x 100	96	94	89	88	91	91	96	97	95	92	89	90	88	89	92	94	96
(2) 1960-1964	11.8	8.6	6.2	8.6	10.8	13.3	20.2	33.1	62.2	111.3	213.7	345.2	518.6	742.5	994.8	1352.0	1713.1
MU/USA x 100	102	95	90	91	96	92	97	97	100	96	97	95	93	93	93	93	96
(3) 1970-1975	6.9	7.2	5.5	7.7	10.6	12.5	18.4	31.6	58.9	122.5	220.4	380.1	608.3	867.1	1161.1	1592.2	1927.3
MU/USA x 100	100	100	100	96	96	95	99	99	95	100	99	99	99	97	98	96	95
(3) ÷ (1) x 100	58	87	96	99	104	91	92	102	105	117	126	125	132	134	131	123	109

MU = Moderately urban
USA = United States

W H I T E   F E M A L E

Mortality rate/100,000

Year	0-4	5-9	10-14	15-19	20-24	25-29	30-34	35-39	40-44	45-49	50-54	55-59	60-64	65-69	70-74	75-84	85+
(4) 1950-1954	10.6	7.1	5.4	6.1	7.2	14.5	28.5	49.4	91.6	144.2	209.2	285.3	389.3	494.7	675.0	955.0	1321.7
MU/USA x 100	100	103	104	98	90	100	103	95	96	96	96	93	94	94	95	95	98
(5) 1960-1964	9.6	6.5	4.7	6.4	7.1	12.1	23.5	47.3	83.7	132.8	201.2	268.1	354.2	463.9	619.6	872.9	1181.3
MU/USA x 100	101	90	85	98	99	100	98	99	97	94	94	95	95	94	96	96	97
(6) 1970-1975	5.5	5.3	4.3	5.0	6.0	10.7	21.0	41.1	73.7	134.6	200.4	293.1	376.7	483.6	607.3	850.5	1128.6
MU/USA x 100	98	98	100	94	95	100	102	102	96	97	96	97	96	97	96	96	97
(6) ÷ (4) x 100	52	75	80	82	83	74	74	83	80	93	96	103	97	98	90	89	85

TABLE A1-32

AGE-SPECIFIC CANCER MORTALITY RATES RURAL COUNTIES, 1950-1975

W H I T E   M A L E

Mortality rate/100,000

Year	0-4	5-9	10-14	15-19	20-24	25-29	30-34	35-39	40-44	45-49	50-54	55-59	60-64	65-69	70-74	75-84	85+
(1) 1950-1954	11.9	8.1	6.2	8.5	11.5	16.0	20.7	28.7	51.2	94.5	159.0	257.3	399.0	554.6	784.2	1237.2	1754.6
RU/USA x 100	96	92	97	96	103	107	99	90	86	84	81	76	76	76	81	90	95
(2) 1960-1964	11.2	7.9	6.5	9.8	13.4	16.1	21.0	33.9	60.5	106.0	193.3	310.8	467.2	663.0	884.4	1278.5	1743.6
RU/USA x 100	97	87	94	103	119	111	101	99	97	91	88	85	83	79	83	88	97
(3) 1970-1975	6.7	6.7	5.1	8.3	11.1	13.2	19.1	33.2	63.5	124.5	219.4	369.9	575.4	819.7	1072.4	1476.7	1889.0
RU/USA x 100	97	93	93	104	101	101	103	104	102	100	99	96	94	92	90	89	93
(3) ÷ (1) x 100	56	83	82	98	97	83	92	116	124	132	138	144	144	148	137	119	108

W H I T E   F E M A L E

Mortality rate/100,000

Year	0-4	5-9	10-14	15-19	20-24	25-29	30-34	35-39	40-44	45-49	50-54	55-59	60-64	65-69	70-74	75-84	85+
(4) 1950-1954	9.7	5.8	4.6	6.2	8.5	15.4	28.1	50.2	86.6	135.6	189.0	262.3	364.2	471.3	656.7	956.4	1329.8
RU/USA x 100	92	84	88	100	106	106	101	96	91	90	87	85	88	89	92	95	99
(5) 1960-1964	8.7	6.3	5.6	6.6	7.4	12.8	24.2	47.2	82.1	129.3	192.9	254.5	329.8	438.9	589.8	869.1	1202.1
RU/USA x 100	92	88	102	102	103	106	101	99	95	91	90	90	89	88	91	95	98
(6) 1970-1975	5.7	5.0	4.0	5.2	7.0	10.9	20.5	39.8	75.5	129.2	192.1	273.3	352.5	450.9	577.5	833.5	1135.3
RU/USA x 100	102	93	93	98	111	102	100	99	98	94	92	91	90	90	92	94	97
(6) ÷ (4) x 100	59	86	87	84	82	71	73	79	87	95	102	104	97	96	88	87	85

RU = Rural
USA = United States

TOTAL POPULATION AGE-ADJUSTED CANCER MORTALITY RATES OF THE WHITE MALE
POPULATION, METROPOLITAN COUNTIES OF THE UNITED STATES,
1950 - 1975

Type	1950 - 1954				1960 - 1964				1970 - 1975			
	Rate	$\frac{ME a}{USA}$	$\frac{ME}{RU}$	$\frac{ME}{SU}$	Rate	$\frac{ME}{USA}$	$\frac{ME}{RU}$	$\frac{ME}{SU}$	Rate	$\frac{ME}{USA}$	$\frac{ME}{RU}$	$\frac{ME}{SU}$
1. Lung	31.5	128	185	105	46.6	115	144	102	61.1	104	111	100
2. Stomach	23.7	114	126	107	15.8	116	131	107	10.5	119	136	111
3. Prostate	18.4	101	102	99	17.1	99	98	99	17.7	100	99	101
4. Large intestine	19.7	125	161	108	19.9	121	151	107	21.8	120	141	110
5. Rectum	11.8	137	200	110	9.3	129	175	109	7.1	125	148	113
6. Pancreas	9.4	115	136	103	10.7	107	115	101	10.9	106	110	103
7. Leukemia	8.5	108	115	102	9.0	99	100	99	8.6	100	99	101
8. Bladder	8.8	129	180	106	8.0	119	151	94	7.7	113	133	104
9. Liver	6.6	108	114	105	5.6	110	119	104	3.4	113	131	103
10. Tongue, mouth	5.9	131	203	105	5.8	129	193	105	4.9	120	158	104
11. Esophagus	6.2	148	270	111	5.4	135	200	110	5.1	124	155	111
12. Lymphoma	4.5	115	136	105	5.5	108	120	104	6.1	107	115	102
13. Brain/cent. nerv.	4.3	110	123	102	4.8	107	112	102	4.9	100	100	96
14. Kidney	4.0	118	143	103	4.3	110	123	102	4.6	105	109	102
15. Larynx	3.4	142	243	106	3.1	124	172	100	3.1	119	141	107
16. Hodgkin's disease	2.4	104	114	100	2.4	104	109	100	1.9	106	106	106
17. Other skin	1.6	84	73	100	1.2	86	71	92	0.7	78	64	88
18. Bone	1.7	100	94	94	1.2	100	100	100	1.0	100	100	100
19. Multiple myeloma	1.4	117	140	100	1.9	100	106	100	2.4	96	96	96
20. Melanoma	1.2	100	120	92	1.6	100	94	94	2.1	95	100	91
21. Total	191.1	117	141	105	195.2	112	129	103	201.2	106	114	102

ME = Metropolitan counties
USA = United States
SU = Strongly urban counties

[a]All rates have been multiplied by 100.

TOTAL POPULATION AGE-ADJUSTED CANCER MORTALITY RATES OF THE WHITE FEMALE POPULATION,
METROPOLITAN COUNTIES OF THE UNITED STATES, 1950 - 1975

	Type	Rate	ME[a]/USA	ME/RU	ME/SU	Rate	ME/USA	ME/RU	ME/SU	Rate	ME/USA	ME/RU	ME/SU
			1 9 5 0 - 1 9 5 4			1 9 6 0 - 1 9 6 4				1 9 7 0 - 1 9 7 5			
1.	Breast	29.6	117	140	105	28.9	114	134	105	29.4	113	129	105
2.	Large intestine	19.4	113	127	105	18.1	112	126	105	16.8	111	117	107
3.	Stomach	12.5	113	119	107	8.0	116	129	110	5.0	116	128	109
4.	Cervix ut.	8.8	93	97	94	6.6	88	80	93	3.9	87	75	95
5.	Corpus ut.	8.0	95	90	100	5.7	104	106	104	4.3	105	105	102
6.	Ovary	9.3	115	141	103	9.5	110	123	104	9.2	107	116	103
7.	Liver	7.1	100	96	103	5.2	104	102	104	2.8	104	100	100
8.	Rectum	7.3	124	162	107	5.4	120	142	108	3.9	118	130	111
9.	Leukemia	5.9	107	116	102	5.9	102	104	100	5.2	100	98	100
10.	Pancreas	6.0	113	128	105	6.4	108	116	103	6.7	108	118	105
11.	Lung	5.4	115	129	104	7.1	116	142	104	14.1	109	136	99
12.	Bladder	3.1	111	135	103	2.5	109	125	104	2.1	111	117	105
13.	Lymphoma	3.0	115	143	103	3.7	109	123	103	4.1	105	114	103
14.	Brain/cent. nervous	3.0	115	136	107	3.2	107	119	103	3.4	100	103	97
15.	Kidney	2.1	111	117	105	2.0	100	105	100	2.1	105	105	105
16.	Hodgkin's disease	1.4	108	127	100	1.5	115	125	107	1.2	109	120	109
17.	Melanoma	0.9	100	100	100	1.1	92	85	100	1.4	100	100	100
18.	Multiple myeloma	0.9	113	129	100	1.4	108	108	100	1.6	94	94	94
19.	Total	151.3	110	120	103	137.9	109	118	104	132.8	107	115	102

ME = Metropolitan counties
USA= United States
SU = Strongly urban counties

[a]All ratios have been multiplied by 100.

TOTAL POPULATION AGE-ADJUSTED CANCER MORTALITY RATES OF THE WHITE MALE POPULATION
NORTHEASTERN, STRONGLY URBAN COUNTIES OF THE UNITED STATES, 1950 - 1975

	1 9 5 0 - 5 4			1 9 6 0 - 6 4			1 9 7 0 - 7 5		
Type	Rate	NESU[a] / USA	NESU / SU	Rate	NESU / USA	NESU / SU	Rate	NESU / USA	NESU / SU
1. Lung	33.4	136	111	47.0	116	103	60.1	102	98
2. Stomach	25.3	122	114	17.4	128	118	11.6	132	122
3. Prostate	18.1	99	98	16.8	97	98	17.4	98	99
4. Large intestine	21.8	139	120	22.3	135	120	23.5	130	119
5. Rectum	13.9	161	130	10.5	146	124	7.8	137	124
6. Pancreas	9.6	117	105	11.0	110	104	11.1	108	105
7. Leukemia	8.4	106	101	9.0	99	99	8.3	97	98
8. Bladder	9.4	138	113	8.4	125	99	8.1	119	109
9. Liver	6.6	108	105	5.8	114	107	3.3	110	100
10. Tongue, mouth	6.4	142	114	5.9	131	107	5.2	127	111
11. Esophagus	6.7	160	120	5.8	145	118	5.2	127	113
12. Lymphoma	4.4	113	102	5.2	102	98	6.0	105	100
13. Brain/ central nervous	4.2	108	100	4.7	104	100	4.6	94	90
14. Kidney	4.1	121	105	4.3	110	102	4.7	107	104
15. Larynx	3.9	163	122	3.5	140	113	3.3	127	114
16. Hodgkin's disease	2.7	117	113	2.6	113	108	2.1	117	117
17. Other skin	1.6	84	100	1.1	79	85	0.7	78	88
18. Bone	1.9	112	106	1.3	108	108	1.0	100	100
19. Multiple myeloma	1.3	108	93	1.8	95	95	2.4	96	96
20. Melanoma	1.2	100	92	1.6	100	95	2.2	100	96
21. Total	202.1	124	111	203.1	117	107	207.3	109	105

NESU   = Northeast strongly urban counties

USA   = United States

SU   = Strongly urban counties

[a]All ratios have been multiplied by 100.

TOTAL POPULATION AGE-ADJUSTED CANCER MORTALITY RATES OF THE WHITE MALE POPULATION, NORTHCENTRAL, STRONGLY URBAN COUNTIES OF THE UNITED STATES, 1950 - 1975

	Type	Rate	NCSU[a]/USA	NCSU/SW	Rate	NCSU/USA	NCSU/SU	Rate	NCSU/USA	NCSU/SU
			1950-54			1960-64			1970-75	
1.	Lung	28.3	115	94	44.6	110	98	61.2	104	100
2.	Stomach	23.1	112	105	15.1	111	103	9.2	105	97
3.	Prostate	19.6	107	106	18.0	104	105	17.4	98	99
4.	Large intestine	18.8	120	103	19.3	117	104	20.2	112	102
5.	Rectum	10.7	124	100	9.0	125	106	6.6	116	105
6.	Pancreas	8.9	109	98	10.6	106	100	10.2	99	96
7.	Leukemia	8.4	106	101	9.2	101	101	8.9	103	105
8.	Bladder	8.2	121	99	8.0	119	94	7.6	112	103
9.	Liver	6.9	113	110	5.6	100	104	3.4	113	103
10.	Tongue, mouth	5.5	122	98	5.7	127	104	4.6	112	98
11.	Esophagus	6.1	145	109	5.3	133	108	5.0	122	109
12.	Lymphoma	4.5	115	105	5.7	112	108	6.2	109	103
13.	Brain/central ner.	4.4	113	105	4.9	109	104	4.6	94	90
14.	Kidney	3.9	115	100	4.6	118	110	4.7	107	104
15.	Larynx	2.9	121	91	3.1	124	100	3.0	115	103
16.	Hodgkin's disease	2.3	100	96	2.4	104	100	1.8	100	100
17.	Other skin	1.5	79	94	1.2	86	92	0.7	78	88
18.	Bone	1.9	112	106	1.2	100	100	1.0	100	100
19.	Multiple myeloma	1.3	108	93	1.9	100	100	2.5	100	100
20.	Melanoma	1.0	83	77	1.4	88	83	1.8	82	78
21.	Total	182.9	112	100	191.1	110	101	199.6	105	101

NCSU = Northcentral strongly urban counties

USA = United States

SU = Strongly urban counties

[a]All ratios have been multiplied by 100.

TOTAL POPULATION AGE-ADJUSTED CANCER MORTALITY RATES OF THE WHITE MALE POPULATION,
SOUTHERN, STRONGLY URBAN COUNTIES OF THE UNITED STATES,
1950 - 1975

Type	1 9 5 0 - 5 4			1 9 6 0 - 6 4			1 9 7 0 - 7 5		
	Rate	SSU[a] / USA	SSU / SU	Rate	SSU / USA	SSU / SU	Rate	SSU / USA	SSU / SU
1. Lung	29.9	122	100	49.4	122	108	67.5	115	110
2. Stomach	15.7	76	71	10.8	79	73	7.4	84	78
3. Prostate	18.3	100	99	16.7	97	97	17.0	96	97
4. Large intestine	13.6	87	75	14.9	90	80	17.1	94	86
5. Rectum	6.5	76	61	5.3	74	62	4.6	81	73
6. Pancreas	8.5	104	93	10.3	103	97	10.6	103	100
7. Leukemia	7.9	100	95	9.2	101	101	8.4	98	99
8. Bladder	7.1	104	86	6.4	96	75	6.6	97	89
9. Liver	6.2	102	98	5.3	104	98	3.3	110	100
10. Tongue, mouth	5.4	120	96	5.7	127	104	4.7	115	100
11. Esophagus	4.0	95	71	4.0	100	82	4.1	100	89
12. Lymphoma	3.9	100	91	4.9	96	92	5.5	96	92
13. Brain/ central nervous	4.0	103	95	4.6	102	98	5.4	110	106
14. Kidney	3.4	100	87	3.6	92	86	4.4	100	98
15. Larynx	3.0	125	94	3.0	120	97	3.0	115	103
16. Hodgkin's disease	2.2	96	92	2.3	100	96	1.7	94	94
17. Other skin	2.3	121	144	1.7	121	131	1.0	111	125
18. Bone	1.6	94	89	1.1	92	92	0.9	90	90
19. Multiple myeloma	1.4	117	100	1.8	95	95	2.4	96	96
20. Melanoma	1.7	142	131	1.9	119	112	2.6	118	113
21. Total	163.6	101	90	178.8	103	95	195.9	103	99

SSU = Southern strongly urban counties

USA = United States

SU = Strongly urban counties

[a]All ratios have been multiplied by 100.

TOTAL POPULATION AGE-ADJUSTED CANCER MORTALITY RATES OF THE WHITE MALE POPULATION,
WESTERN, STRONGLY URBAN COUNTIES OF THE UNITED STATES, 1950 - 1975

	Type	Rate	WSU[a]/USA	WSU/SU	Rate	WSU/USA	WSU/SU	Rate	WSU/USA	WSU/SU
		1 9 5 0 - 5 4			**1 9 6 0 - 6 4**			**1 9 7 0 - 7 5**		
1.	Lung	25.8	105	86	41.2	101	90	56.3	96	92
2.	Stomach	19.6	95	89	13.1	96	89	9.0	102	95
3.	Prostate	17.8	97	96	17.1	99	99	17.9	101	102
4.	Large intestine	13.8	88	76	14.5	88	78	16.4	91	83
5.	Rectum	8.0	93	75	7.0	97	82	5.1	89	81
6.	Pancreas	8.9	109	98	10.0	100	94	10.3	100	97
7.	Leukemia	8.3	106	100	9.0	99	99	8.4	98	99
8.	Bladder	7.3	107	88	6.9	103	81	6.6	97	89
9.	Liver	5.1	84	81	4.4	86	81	3.1	103	94
10.	Tongue, mouth	4.4	98	79	4.4	98	80	3.9	95	83
11.	Esophagus	3.5	83	63	3.8	95	78	3.8	93	83
12.	Lymphoma	4.4	113	102	5.5	108	104	5.5	96	92
13.	Brain/central nervous	4.2	108	100	4.7	104	100	5.4	110	106
14.	Kidney	3.5	103	90	3.9	100	93	4.1	93	91
15.	Larynx	2.3	96	72	2.3	92	74	2.2	85	76
16.	Hodgkin's disease	2.1	91	88	2.1	91	88	1.7	94	94
17.	Other skin	1.6	84	100	1.2	86	92	1.0	111	125
18.	Bone	1.4	82	78	0.9	75	75	0.9	90	90
19.	Multiple myeloma	1.6	133	114	2.2	116	116	2.4	96	96
20.	Melanoma	1.4	117	108	1.8	113	106	2.6	118	113
21.	Total	159.3	98	87	171.1	98	90	184.1	97	93

WSU = Western strongly urban counties

US = United States

SU = Strongly urban counties

[a] All ratios have been multiplied by 100.

TOTAL POPULATION AGE-ADJUSTED CANCER MORTALITY RATES OF THE WHITE FEMALE POPULATION, NORTHEASTERN, STRONGLY URBAN COUNTIES OF THE UNITED STATES,1950-1975

Type	1950-1954			1960-1964			1970-1975		
	Rate	NESU[a] / USA	NESU / SU	Rate	NESU / USA	NESU / SU	Rate	NESU / USA	NESU / SU
1. Breast	31.0	122	110	30.3	120	110	30.8	118	110
2. Large intestine	21.9	128	119	20.2	125	117	18.0	118	115
3. Stomach	13.9	125	119	8.8	128	121	5.7	133	124
4. Cervix uteri	8.0	84	85	5.6	75	79	3.7	82	90
5. Corpus uteri	8.3	99	104	6.0	109	109	4.5	110	107
6. Ovary	9.5	117	106	9.5	110	104	9.4	109	106
7. Liver	7.3	103	106	5.3	106	106	2.9	107	104
8. Rectum	8.3	141	122	6.0	133	120	4.4	133	126
9. Leukemia	6.0	109	103	6.0	103	102	5.1	98	98
10. Pancreas	6.2	117	109	6.7	114	108	6.9	111	108
11. Lung	5.5	117	106	7.0	115	103	13.7	106	96
12. Bladder	3.3	117	110	2.6	113	108	2.2	116	110
13. Lymphoma	2.9	112	100	3.6	106	100	4.0	103	100
14. Brain/ central ner.	2.8	108	100	3.0	100	97	3.2	94	91
15. Kidney	2.1	111	105	2.1	105	105	2.0	100	100
16. Hodgkin's disease	1.5	115	107	1.6	123	114	1.4	127	127
17. Melanoma	0.8	89	89	1.1	92	100	1.4	100	100
18. Multiple myeloma	1.0	125	111	1.4	108	100	1.6	94	94
19. Total	158.5	115	108	143.6	113	108	137.8	111	106

NESU= Northeast strongly urban counties

USA = United States

SU= Strongly urban counties

[a] All ratios have been multiplied by 100.

TABLE A1-40

TOTAL POPULATION AGE-ADJUSTED CANCER MORTALITY RATES OF THE WHITE FEMALE POPULATION, NORTHCENTRAL, STRONGLY URBAN COUNTIES OF THE UNITED STATES, 1950-1975

	1950-1954			1960-1964			1970-1975		
Type	Rate	NCSU[a] USA	NCSU SU	Rate	NCSU USA	NCSU SU	Rate	NCSU USA	NCSU SU
1. Breast	28.7	113	102	28.1	111	102	28.0	108	100
2. Large intestine	18.4	108	100	17.7	110	103	15.8	104	101
3. Stomach	11.9	107	102	7.2	104	99	4.3	100	93
4. Cervix uteri	9.9	104	105	7.7	103	108	4.4	98	107
5. Corpus uteri	8.8	105	110	6.0	109	109	4.4	110	107
6. Ovary	9.5	117	106	9.9	115	109	9.3	108	104
7. Liver	7.8	110	113	5.7	114	114	3.0	111	107
8. Rectum	6.5	110	96	5.0	111	100	3.6	109	103
9. Leukemia	5.8	105	100	5.7	98	97	5.4	104	104
10. Pancreas	5.6	106	98	6.1	103	98	6.2	100	97
11. Lung	5.2	111	100	6.4	105	94	13.3	103	93
12. Bladder	3.0	107	100	2.4	104	100	2.0	105	100
13. Lymphoma	3.0	115	103	3.8	112	106	4.2	108	105
14. Brain/ central ner.	3.1	119	111	3.2	107	103	3.5	103	100
15. Kidney	2.2	116	110	2.2	110	110	2.2	110	110
16. Hodgkin's disease	1.4	108	100	1.4	108	100	1.1	100	100
17. Melanoma	0.8	89	89	1.0	83	91	1.2	86	86
18. Multiple myeloma	0.9	113	100	1.4	108	100	1.7	100	100
19. Total	148.5	108	102	134.6	107	101	129.3	104	100

NCSU= Northcentral strongly urban counties

USA = United States

SU=Strongly urban counties

[a]All ratios have been multiplied by 100.

239

TOTAL POPULATION AGE-ADJUSTED CANCER MORTALITY RATES OF THE WHITE FEMALE POPULATION, SOUTHERN, STRONGLY URBAN COUNTIES OF THE UNITED STATES, 1950-1975

	1950-1954			1960-1964			1970-1975		
Type	Rate	SSU[a] USA	SSU SU	Rate	SSU USA	SSU SU	Rate	SSU USA	SSU SU
1. Breast	22.8	90	81	23.3	92	85	24.7	95	89
2. Large intestine	14.5	85	79	14.4	89	84	14.1	93	90
3. Stomach	8.1	73	69	5.5	80	75	3.7	86	80
4. Cervix uteri	11.4	120	121	8.3	111	117	4.6	102	112
5. Corpus uteri	7.5	89	94	4.8	87	87	3.5	85	83
6. Ovary	7.6	94	84	8.0	93	88	8.0	93	90
7. Liver	5.7	80	83	4.0	80	80	2.3	85	82
8. Rectum	4.9	83	72	3.7	82	74	2.7	82	77
9. Leukemia	5.3	96	91	5.8	100	98	5.2	100	100
10.Pancreas	5.1	96	89	5.8	98	94	6.2	100	97
11.Lung	4.8	102	92	6.9	113	101	15.4	119	108
12.Bladder	2.8	100	93	2.4	104	100	1.9	100	95
13.Lymphoma	2.6	100	90	3.5	103	97	3.7	95	93
14.Brain /central nerv.	2.6	100	93	3.2	107	103	3.8	112	109
15.Kidney	1.9	100	95	1.8	90	90	1.9	95	95
16.Hodgkin's disease	1.3	100	93	1.2	92	86	1.0	91	91
17.Melanoma	1.1	122	122	1.2	100	109	1.5	107	107
18.Multiple myeloma	0.9	113	100	1.4	108	100	1.6	94	94
19.Total	129.3	94	88	121.1	95	91	122.0	98	94

SSU= Southern strongly urban counties

USA= United States

SU= Strongly urban counties

[a]All ratios have been multiplied by 100.

TOTAL POPULATION AGE-ADJUSTED CANCER MORTALITY RATES OF THE WHITE FEMALE POPULATION, WESTERN, STRONGLY URBAN COUNTIES OF THE UNITED STATES,1950-1975

Type	1950-1954 Rate	WSU[a] USA	WSU SU	1960-1964 Rate	WSU USA	WSU SU	1970-1975 Rate	WSU USA	WSU SU
1. Breast	26.2	103	93	25.7	102	93	26.6	102	95
2. Large intestine	14.6	85	79	13.9	86	81	13.5	89	86
3. Stomach	10.0	90	85	6.6	96	90	4.1	95	89
4. Cervix uteri	9.7	102	103	7.6	101	107	4.1	91	100
5. Corpus uteri	6.7	80	84	4.4	80	80	4.0	98	95
6. Ovary	8.7	107	97	8.4	98	92	8.9	103	100
7. Liver	5.6	79	81	4.3	86	86	2.6	96	93
8. Rectum	5.6	95	82	4.4	98	88	3.0	91	86
9. Leukemia	6.0	109	103	5.9	102	100	5.1	98	98
10.Pancreas	5.3	100	93	5.9	100	95	6.4	103	100
11.Lung	4.7	100	90	7.0	115	103	15.3	119	107
12.Bladder	2.6	93	87	2.2	96	92	2.0	105	100
13.Lymphoma	2.9	112	100	3.7	109	103	4.2	108	105
14.Brain /central ner.	2.9	112	104	3.1	103	100	3.5	103	100
15.Kidney	1.8	95	90	1.8	90	90	1.9	95	95
16.Hodgkin's disease	1.2	92	86	1.3	100	93	0.9	82	82
17.Melanoma	1.1	122	122	1.2	100	109	1.5	107	107
18.Multiple myeloma	0.9	113	100	1.4	108	100	1.9	112	112
19.Total	132.6	96	91	123.9	98	93	125.7	101	97

WSU= Western strongly urban counties

USA= United States
SU= Strongly urban counties

[a]All ratios have been multiplied by 100.

AGE-SPECIFIC CANCER MORTALITY RATES,1950-1975, NORTHEAST STRONGLY URBAN COUNTIES
Mortality Rate/100,000

WHITE MALE

Age group	(1) 1950-54	(2) $(1)^a$ USA	(3) 1960-64	(4) $\frac{(3)}{USA}$	(5) 1970-75	(6) $\frac{(5)}{USA}$	(7) (5)/(1)
0-4	12.5	101	12.4	107	7.2	104	58
5-9	9.2	105	9.7	107	7.3	101	79
10-14	6.5	102	7.5	109	5.4	98	83
15-19	11.0	124	10.1	106	9.1	114	83
20-24	12.6	113	11.9	105	11.6	105	92
25-29	16.5	110	16.3	112	13.6	104	82
30-34	23.4	112	21.0	101	19.0	102	81
35-39	36.8	115	35.5	104	30.8	96	84
40-44	69.2	117	64.3	103	63.3	102	91
45-49	133.5	118	127.7	110	127.4	103	95
50-54	238.8	121	247.6	112	223.7	100	94
55-59	435.1	129	411.1	113	400.8	104	92
60-64	671.8	129	655.7	117	647.0	105	96
65-69	961.5	132	959.9	120	975.6	109	101
70-74	1233.4	128	1289.9	121	1348.5	113	109
75-84	1664.2	121	1744.2	121	1912.1	115	115
85 +	2058.9	112	2015.4	113	2340.9	115	114

WHITE FEMALE

Age group	(1)	(2)	(3)	(4)	(5)	(6)	(7)
0-4	10.8	102	9.9	104	5.4	96	50
5-9	8.2	119	8.4	117	5.5	102	67
10-14	5.4	104	5.6	102	4.4	102	81
15-19	6.9	111	7.0	108	5.9	111	86
20-24	8.5	106	7.9	110	6.6	105	78
25-29	14.7	101	12.2	101	11.3	106	77
30-34	28.8	104	25.2	105	21.1	103	73
35-39	56.4	108	50.5	106	41.3	102	73
40-44	104.1	110	92.9	107	79.6	104	76
45-49	167.9	111	159.9	113	149.8	108	89
50-54	247.6	114	239.3	112	229.8	110	93
55-59	367.2	119	318.9	113	325.5	108	89
60-64	488.9	118	428.1	115	433.5	110	89
65-69	620.8	117	583.7	118	559.2	112	90
70-74	833.0	117	749.7	116	715.7	114	86
75-84	1160.1	115	1024.2	112	1015.6	115	88
85 +	1455.5	108	1340.9	110	1277.8	110	88

[a]All ratios have been multiplied by 100.

TABLE A1-44

AGE-SPECIFIC CANCER MORTALITY RATES, 1950-1975, NORTHCENTRAL STRONGLY URBAN COUNTIES

Mortality Rate/100,000

WHITE MALE

Age group	(1) 1950-54	(2) (1)[a] USA	(3) 1960-64	(4) (3) USA	(5) 1970-75	(6) (5) USA	(7) (5)/(1)
0-4	13.6	110	11.6	100	6.9	100	51
5-9	9.2	105	10.1	111	7.1	99	77
10-14	7.7	120	7.5	109	5.7	104	74
15-19	9.5	107	10.5	111	8.0	100	84
20-24	11.1	99	11.4	101	11.8	107	106
25-29	14.3	95	13.5	93	12.9	98	90
30-34	20.0	96	20.6	96	17.9	96	90
35-39	32.6	102	34.5	101	32.0	100	98
40-44	63.2	107	64.1	103	61.7	100	98
45-49	123.8	110	123.7	106	124.5	100	101
50-54	214.2	109	232.4	105	228.1	102	106
55-59	375.3	111	392.1	108	400.6	104	107
60-64	600.6	115	604.6	108	638.9	104	106
65-69	862.1	118	897.5	112	951.5	107	110
70-74	1125.4	117	1225.7	115	1263.2	106	112
75-84	1524.9	111	1625.0	112	1790.8	108	117
85 +	1928.8	105	1880.1	105	2159.2	106	112

WHITE FEMALE

Age group	(1)	(2)	(3)	(4)	(5)	(6)	(7)
0-4	11.3	107	9.3	98	5.5	98	49
5-9	7.0	101	7.4	103	5.2	96	74
10-14	5.7	110	6.1	111	4.2	98	74
15-19	6.4	103	5.8	89	5.2	98	81
20-24	8.3	104	6.9	96	6.1	97	73
25-29	14.5	100	12.8	106	10.9	102	75
30-34	28.9	104	23.8	99	21.0	102	73
35-39	53.7	103	48.9	102	39.9	99	74
40-44	100.3	106	88.1	102	80.3	105	80
45-49	162.3	108	150.9	106	139.9	101	86
50-54	233.7	107	223.2	105	216.9	104	93
55-59	334.5	109	295.3	105	310.2	103	93
60-64	451.5	109	402.0	108	412.4	105	91
65-69	587.0	111	537.5	108	530.4	106	90
70-74	770.6	108	705.0	109	659.1	105	86
75-84	1078.2	107	964.6	106	920.0	104	85
85 +	1383.8	103	1272.4	104	1190.3	102	86

[a]All ratios have been multiplied by 100.

243

AGE-SPECIFIC CANCER MORTALITY RATES,1950-1975, SOUTHERN STRONGLY URBAN COUNTIES

Mortality Rate/100,000

WHITE MALE

Age group	(1) 1950-54	(2) (1)[a] USA	(3) 1960-64	(4) (3) USA	(5) 1970-75	(6) (5) USA	(7) (5)/(1)
0-4	12.5	101	11.3	97	6.6	96	53
5-9	9.0	102	9.4	103	7.0	97	78
10-14	6.1	95	7.0	101	6.0	109	98
15-19	8.8	99	7.6	80	7.0	88	80
20-24	10.0	89	9.4	83	9.6	87	96
25-29	14.2	95	12.6	87	13.7	105	96
30-34	19.3	92	22.4	108	19.3	104	100
35-39	32.5	102	36.6	107	32.5	102	100
40-44	60.2	102	66.0	106	66.0	106	110
45-49	117.0	103	124.1	107	133.5	108	114
50-54	212.1	108	238.5	108	244.4	110	115
55-59	349.9	103	402.2	110	413.1	107	118
60-64	542.5	104	607.0	108	650.4	106	120
65-69	719.0	99	829.9	104	928.0	104	129
70-74	944.0	98	1083.9	102	1197.2	101	127
75-84	1368.2	99	1397.1	96	1676.7	101	123
85 +	1772.3	96	1658.2	93	1980.4	97	112

WHITE FEMALE

Age group	(1)	(2)	(3)	(4)	(5)	(6)	(7)
0-4	10.1	95	10.1	106	6.3	113	62
5-9	7.0	101	7.8	108	6.0	111	86
10-14	4.8	92	5.5	100	4.5	105	94
15-19	4.9	79	6.1	94	5.7	108	116
20-24	7.4	93	6.8	94	5.9	94	80
25-29	14.1	97	11.2	93	10.4	97	74
30-34	25.5	92	23.4	98	19.8	97	78
35-39	51.6	99	46.5	97	41.0	102	79
40-44	93.8	99	88.5	102	76.6	100	82
45-49	149.7	99	137.7	97	139.5	101	93
50-54	214.1	98	206.8	97	207.2	99	97
55-59	288.7	94	270.9	96	302.5	100	105
60-64	390.5	94	347.4	94	392.5	100	101
65-69	482.2	91	464.2	94	475.5	95	99
70-74	660.2	93	597.4	92	609.0	97	92
75-84	923.2	92	867.1	95	837.7	95	91
85 +	1182.9	88	1183.9	97	1145.8	98	97

[a]All ratios have been multiplied by 100.

AGE–SPECIFIC CANCER MORTALITY RATES,1950-1975,WESTERN STRONGLY URBAN COUNTIES

Mortality Rate/100,000

WHITE MALE

Age group	(1) 1950-54	(2) (1)[a] USA	(3) 1960-64	(4) (3) USA	(5) 1970-75	(6) (5) USA	(7) (5)/(1)
0-4	12.4	100	11.4	98	7.0	101	56
5-9	9.8	111	10.0	110	7.9	110	81
10-14	7.3	114	7.2	104	5.9	107	81
15-19	8.5	96	10.1	106	7.9	99	93
20-24	11.1	99	10.1	89	11.1	101	100
25-29	14.8	99	14.2	98	12.8	98	86
30-34	21.0	101	20.0	96	17.3	93	82
35-39	31.0	97	31.6	92	31.4	98	101
40-44	57.5	97	58.4	93	58.3	94	101
45-49	112.6	100	110.4	95	111.4	90	99
50-54	201.1	102	209.0	95	202.5	91	101
55-59	332.1	98	353.4	97	358.8	93	108
60-64	503.1	96	561.4	100	591.0	96	117
65-69	723.7	99	792.9	99	866.7	97	120
70-74	946.1	98	1043.6	98	1179.4	99	‵125
75-84	1314.5	96	1441.1	100	1651.9	100	126
85 +	1788.5	97	1773.5	99	2066.9	102	116

WHITE FEMALE

Age group	(1)	(2)	(3)	(4)	(5)	(6)	(7)
0-4	12.5	118	9.9	104	5.5	98	44
5-9	7.7	112	7.9	110	5.7	106	74
10-14	6.2	119	5.6	102	4.3	100	69
15-19	6.0	97	6.9	106	5.1	96	85
20-24	7.6	95	7.0	97	5.7	90	75
25-29	12.9	89	10.9	90	9.8	92	76
30-34	25.2	91	23.2	97	19.4	95	77
35-39	50.2	96	45.5	95	38.5	96	77
40-44	94.6	100	85.6	99	74.8	98	79
45-49	147.0	97	139.9	99	138.0	100	94
50-54	213.5	98	220.8	104	216.6	103	101
55-59	309.0	100	282.9	101	315.4	105	102
60-64	399.0	96	362.3	98	414.1	105	104
65-69	507.1	96	487.8	98	520.1	104	103
70-74	658.9	92	606.3	94	625.5	95	95
75-84	941.6	93	870.6	95	859.1	97	91
85 +	1275.5	95	1159.6	95	1117.9	96	88

[a]All ratios have been multiplied by 100.

TABLE A1-47

AGE-ADJUSTED CANCER MORTALITY RATES OF THE WHITE MALE
POPULATION, CENTRAL CITY COUNTIES OF THE UNITED
STATES, 1950-1975

		1950-1954				1960-1964				1970-1975		
Type	Rate	CC[a]/USA	CC/RU	CC/SU	Rate	CC/USA	CC/RU	CC/SU	Rate	CC/USA	CC/RU	CC/SU
1. Lung	34.1	139	201	125	48.9	120	151	113	62.0	105	113	103
2. Stomach	26.0	126	138	124	17.2	126	142	120	11.1	126	144	117
3. Prostate	18.5	101	102	99	17.2	99	98	99	17.5	99	98	97
4. Large intestine	21.0	134	172	112	21.5	130	163	113	22.1	122	143	105
5. Rectum	13.2	153	224	127	10.1	140	191	119	7.3	128	152	109
6. Pancreas	9.9	121	143	115	11.2	112	120	111	11.1	108	112	105
7. Leukemia	8.7	110	118	112	9.3	102	103	106	8.6	100	99	100
8. Bladder	9.3	137	190	115	8.4	125	158	109	7.8	115	134	101
9. Liver	7.1	116	122	116	6.1	120	130	120	3.6	120	138	120
10. Tongue, mouth	6.5	144	224	125	6.5	144	217	127	5.4	132	174	123
11. Esophagus	7.3	174	317	149	6.2	155	230	135	5.5	134	167	122
12. Lymphoma	4.6	118	139	112	5.7	112	124	112	6.2	109	117	105
13. Brain/central ner.	4.5	115	129	118	4.9	109	114	107	4.8	98	98	98
14. Kidney	4.2	124	150	114	4.5	115	129	107	4.6	105	110	100
15. Larynx	3.9	163	279	150	3.6	144	200	133	3.5	135	159	130
16. Hodgkin's disease	2.5	109	119	109	2.5	109	114	109	2.0	111	111	105
17. Other skin	1.5	79	68	94	1.2	86	71	109	0.7	78	64	100
18. Bone	1.8	106	100	100	1.3	108	108	108	1.0	100	100	100
19. Multiple myeloma	1.4	117	140	108	1.9	100	106	106	2.4	96	96	100
20. Melanoma	1.2	100	120	100	1.6	100	94	100	2.0	91	95	87
Total	204.2	126	151	117	206.5	119	136	112	208.9	110	119	105

USA=United States; CC=central city counties; SU=suburban counties; RU=rural counties

[a] All ratios have been multiplied by 100.

246

AGE-ADJUSTED CANCER MORTALITY RATES OF THE WHITE MALE POPULATION,
SUBURBAN COUNTIES OF THE UNITED STATES, 1950-1975

	Type	1950-1954 Rate	$\frac{SU^a}{USA}$	1960-1964 Rate	$\frac{SU}{USA}$	1970-1975 Rate	$\frac{SU}{USA}$
1.	Lung	27.3	111	43.4	107	60.1	102
2.	Stomach	20.9	101	14.3	105	9.5	108
3.	Prostate	18.7	102	17.3	100	18.0	102
4.	Large intestine	18.7	119	19.0	115	21.1	117
5.	Rectum	10.4	121	8.5	118	6.7	118
6.	Pacreas	8.6	105	10.1	101	10.6	103
7.	Leukemia	7.8	99	8.8	97	8.6	100
8.	Bladder	8.1	119	7.7	115	7.7	113
9.	Liver	6.1	100	5.1	100	3.0	100
10.	Tongue, mouth	5.2	116	5.1	113	4.4	107
11.	Esophagus	4.9	117	4.6	115	4.5	110
12.	Lymphoma	4.1	105	5.1	100	5.9	104
13.	Brain/central nerv.	3.8	97	4.6	102	4.9	100
14.	Kidney	3.7	109	4.2	108	4.6	105
15.	Larynx	2.6	108	2.7	108	2.7	104
16.	Hodgkin's disease	2.3	100	2.3	100	1.9	106
17.	Other skin	1.6	84	1.1	79	0.7	78
18.	Bone	1.8	106	1.2	100	1.0	100
19.	Multiple myeloma	1.3	108	1.8	95	2.4	96
20.	Melanoma	1.2	100	1.6	100	2.3	105
	Total	174.1	107	183.9	106	198.6	105

[a] All ratios have been multiplied by 100.

USA= United States
SU = Suburban counties

TABLE A1-49

AGE-ADJUSTED CANCER MORTALITY RATES OF THE WHITE FEMALE
POPULATION, CENTRAL CITY COUNTIES OF THE UNITED STATES, 1950-1975

Type	1950-1954				1960-1964				1970-1975			
	Rate	$\frac{CC^a}{USA}$	$\frac{CC}{RU}$	$\frac{CC}{SU}$	Rate	$\frac{CC}{USA}$	$\frac{CC}{RU}$	$\frac{CC}{SU}$	Rate	$\frac{CC}{USA}$	$\frac{CC}{RU}$	$\frac{CC}{SU}$
1. Breast	30.6	120	144	108	30.1	119	139	108	30.2	116	132	104
2. Large intestine	20.3	119	133	106	19.0	118	132	104	16.7	110	117	99
3. Stomach	13.8	124	131	127	8.7	126	140	123	5.4	126	138	115
4. Cervix uteri	8.8	93	97	104	6.6	88	80	106	4.0	89	77	108
5. Corpus uteri	8.1	96	91	98	6.0	109	111	105	4.5	110	110	107
6. Ovary	9.7	120	147	111	9.9	115	129	106	9.3	108	118	101
7. Liver	7.6	107	103	117	5.6	112	110	117	3.0	111	107	107
8. Rectum	8.0	136	178	125	5.7	127	150	112	4.0	121	133	105
9. Leukemia	6.1	111	120	111	6.2	107	109	111	5.3	102	100	106
10. Pancreas	6.3	119	134	117	6.8	115	124	115	6.9	111	121	106
11. Lung	5.9	126	140	123	7.3	120	146	114	14.2	110	137	101
12. Bladder	3.2	114	139	103	2.5	109	125	93	2.1	111	117	105
13. Lymphoma	3.1	119	148	107	3.8	112	127	106	4.2	108	117	108
14. Brain/central ner.	3.1	119	141	119	3.3	110	122	114	3.4	100	103	100
15. Kidney	2.3	121	128	121	2.2	110	116	110	2.1	105	105	100
16. Hodgkin's disease	1.5	115	136	107	1.6	123	133	114	1.3	118	130	118
17. Melanoma	0.8	89	89	89	1.1	92	85	92	1.3	93	108	87
18. Multiple myeloma	1.0	124	143	111	1.4	108	108	100	1.6	94	94	94
Total	158.5	115	126	111	143.9	113	123	108	136.6	110	118	103

RU= Rural counties; SU= Suburban counties;   USA=United States; CC=central city counties;
[a]All ratios have been multiplied by 100.

TABLE A1–50

AGE-ADJUSTED CANCER MORTALITY RATES OF THE WHITE FEMALE POPULATION,
SUBURBAN COUNTIES OF THE UNITED STATES, 1950-1975

	Type	1950-1954 Rate	1950-1954 SU/USA	1960-1964 Rate	1960-1964 SU/USA	1970-1975 Rate	1970-1975 SU/USA
1.	Breast	28.3	111	27.9	110	29.1	112
2.	Large intestine	19.2	112	18.2	113	16.9	111
3.	Stomach	10.9	98	7.1	103	4.7	109
4.	Cervix uteri	8.5	89	6.2	83	3.7	82
5.	Corpus uteri	8.3	99	5.7	104	4.2	102
6.	Ovary	8.7	107	9.3	108	9.2	107
7.	Liver	6.5	92	4.8	96	2.8	104
8.	Rectum	6.4	108	5.1	113	3.8	115
9.	Leukemia	5.5	100	5.6	97	5.0	96
10.	Pancreas	5.4	102	5.9	100	6.5	105
11.	Lung	4.8	102	6.4	105	14.1	109
12.	Bladder	3.1	111	2.7	117	2.0	105
13.	Lymphoma	2.9	112	3.6	106	3.9	100
14.	Brain/central ner.	2.6	100	2.9	97	3.4	100
15.	Kidney	1.9	100	2.0	100	2.1	105
16.	Hodgkin's disease	1.4	108	1.4	108	1.1	100
17.	Melanoma	0.9	100	1.2	100	1.5	107
18.	Multiple myeloma	0.9	113	1.4	108	1.7	100
	Total	142.4	103	132.7	105	132.5	106

[a]All ratios have been multiplied by 100.

USA=United States

SU = Suburban counties

249

2 Selected Statistical Methods

This appendix reviews rank correlation for trend analysis and discriminant analysis.

RANK CORRELATION FOR TREND ANALYSIS

Rank correlation analysis is widely used. It is reviewed here only because it is used in an unusual manner: to compare trends in cancer mortality rates. The author had intended to use parametric methods to correlate time change indices between 1950–54 and 1970–75. High correlations and low standard errors of estimate imply parallel change between the different populations; low correlations and high standard errors imply the converse. As an example, let us assume that one wished to compare the time change indices in Region I and Region II, as shown by the sample data in the following table.

Unweighted and weighted correlations were computed between the time change indices in columns (3) and (6). The unweighted correlation was .966 (significant at .01). To reduce the chances that the low rate diseases are controlling the results, the square root of the average of the 1950–54 and 1970–75 rates were used to weight the Region 1 time change indices. The square root weighting scheme was chosen for this illustration because it represents a compromise between no weighting and weighting both indices by total population. The results were not significantly changed by weighting (correlation dropped from .966 to .946).

The initial pair of correlations between the two time change indices are misleading. The strong association is dependent upon disease A, which has the highest rates and the highest increase in rates. If the cor-

Disease	Region I Cancer Mortality Rate/100,000			Region II Cancer Mortality Rate/100,000		
	Time Period		Time Change Index ¾ × 100	Time Period		Time Change Index ¾ × 100
	1	2		1	2	
A	50	150	300	30	140	467
B	30	40	133	25	33	132
C	20	15	75	18	16	89
D	10	8	80	11	6	55
E	6	3	50	5	2.5	50
F	2	2	100	1.8	2.1	117
G	1	1.2	120	.8	1.0	125
H	.8	.5	63	.9	1.1	122
I	.7	.9	129	.7	.9	129
J	.7	.3	43	.5	.1	20

relations are recomputed without disease A, they drop from .966 to .804 for the unweighted and from .946 to .336 for the weighted correlations (see summary below). In short, the emphasis placed on extreme values by the parametric correlation procedure may overly influence a relationship and mislead the analyst.

Unfortunately, the sample data closely resemble the actual data. Disease A in the sample data is lung cancer in the actual data. Lung cancer has the highest rate and the highest time change index. It dominated many of the test relationships.

Spearman rank correlation, therefore, was used. The Spearman rank results are not as strongly influenced by the extreme values as the parametric correlations. The nine-disease correlation case is much closer to the ten-disease correlation case (.88 versus .92) than are the parametric correlations as compared in the following table.

Overall, the unusual circumstances of one disease exhibiting the highest rate and highest rate of change necessitated the use of nonparametric correlation methods.

DISCRIMINANT ANALYSIS

Discriminant analysis is a method for testing and generating hypotheses when one dimension of the data is categorical (Johnston, 1978; Nie et al.,

Pearson Correlation

Number of Diseases	Weighted By Square Root of Rate _r_	Sign	Unweighted _r_	Sign	Spearman Correlation	Sign
10	.946	.01	.966	.01	.92	.01
9 (without disease A)	.336	≤.20	.804	.01	.88	.01

1975; Van DeGeer, 1971). For example, one might hypothesize that the urban areas of Northeast United States have higher cancer mortality rates than the Southern and Western urban areas because the Northeast was the cradle of urbanization in the United States. If only a few diseases and regions are of interest, a difference of means test can be used to evaluate the hypotheses. However, when many diseases and regions are of interest and the diseases have similar geographical patterns, discriminant analysis becomes useful because it can simultaneously evaluate multiple diseases and multiple regions. The results will identify those groups of diseases, if any, which most clearly differentiate the regions from one another.

Seven pieces of output are particularly important in the context of this volume. The first is the *mean values* for each group and the *grand mean*. These data enable the analyst to preview the most important differences between the regions. For example, in the same data below, region B has the highest values for three of the four diseases, region C has the lowest mean values for three of the four.

Regions
Rate/(100,000)

Disease	A	B	C	Grand mean
1	10	12	8	10
2	6	5	4	5
3	20	30	25	25
.	.	.	.	.
.	.	.	.	.
.	.	.	.	.
15	40	45	35	40

The second important set of data are *pooled within-group bivariate correlations* between the diseases. Diseases 1, 2, and 3 are strongly correlated with one another, but not with disease 15. It is likely that diseases 1, 2, and 3 will be grouped together in the multivariate analyses. See the table below

Diseases	1	2	3	· · · ·	15
1		.75	.60	· · · ·	.00
2			.65	· · · ·	.15
3				· · · ·	−.10
·					·
·					·
·					·
15					

A third part of the evaluation is the criteria used to determine which diseases most strongly differentiate the groups. *Wilks' lambda* is the measure most commonly used. The smaller the value of lambda, the more discriminating the disease. Lambda is transformed into a chi-square value to obtain a significance value for each disease. With respect to the sample data, disease 2 with the lowest lambda and highest F value has the highest significance. Disease 3 is close to disease 2 in discriminating power. Diseases 1 and 15 are weaker discriminators. The following table shows the evaluation criteria.

Disease	Wilks' Lambda	F	Significance
1	.70	2.10	.050
2	.30	10.50	.001
3	.45	8.30	.001
·	·	·	·
·	·	·	·
·	·	·	·
15	.98	.50	.450

The *stepwise* process is used in this volume. Using lambda and the significance values as the evaluation criteria, the stepwise process begins by incorporating the most discriminating disease into the solution. Then controlling for the first disease, the second most significant disease is brought into the solution. This step-by-step process is continued until the remaining diseases add no further significant discriminating power to the solution.

With respect to the sample data, let us assume that diseases 2, 3, and 15 have been incorporated into the stepwise solution as below:

Disease	Wilks' Lambda	Significance
2	.30	.0010
3	.20	.0015
15	.10	.0002

In the process of choosing which diseases are significant and which are not, multiple disease *discriminant functions* are calculated. In the three region sample case (regions A, B, and C), up to two discriminant functions are calculated. Let us assume that two functions were developed.

Rotated Correlations between Discriminant Functions and Discrimination Variables

1	.20	.15
2	.80	.12
3	.70	.05
.	.	.
.	.	.
.	.	.
15	.01	.55
Cumulative percentage of variation explained	80	20

The first function is strongly associated with diseases 2 and 3 and explains 80 percent of the variance. The second function identifies with disease 15 and explains 20 percent of the variance.

The sixth piece of information is the *group means* of the regions evaluated for each discriminant function. In the sample case, below, area A is strongly differentiated from area C by the first function (Area A = 6.0; Area C = −4.0). The second function differentiates region B from C and to a lesser extent A.

	Function	
Region	1	2
A	6.0	−1.0
B	1.0	3.0
C	−4.0	−2.5

The last piece of information is the classification results. The actual region of each of the 15 regions is compared to the region predicted by the discriminant function; see below.

Area	Actual Region	Highest Group (probability)		Second Highest Group (probability)	
1	A	A	1.00	B	.00
2	A	A	1.00	B	.00
3	A	C	.75	A	.25
4	A	A	.95	C	.05
5	A	A	.80	C	.20
6	B	B	.75	A	.25
7	B	B	.55	A	.45
8	B	A	.90	C	.10
9	B	B	1.00	C	.00
10	B	B	1.00	A	.00
11	C	C	1.00	B	.00
12	C	C	.98	A	.02
13	C	C	.60	A	.40
14	C	C	.55	B	.45
15	C	C	.95	A	.05

Thirteen of the 15 sample areas have been classified in the actual region, two were misclassified. Observation 3 has a 75 percent probability of being in region C and only a 25 percent probability of being in region A. Observation 8 belongs in region B, but has a 90 percent probability of being in region A.

REFERENCES

Johnston, R. 1978. *Multivariate Statistical Analysis in Geography.* New York: Longman.

Nie, N., Hull, C., Jenkins, J., Steinbrenner, K., and Bent, D. 1975. *SPSS: Statistical Package for the Social Sciences,* 2nd ed. New York: McGraw-Hill.

Van De Geer, J. 1971. *Introduction to Multivariate Analysis for the Social Sciences,* San Francisco: Freeman.

3 Cancer Mortality Trends 1930–75 and Cancer Incidence Trends 1937–76

It is not essential to the thesis of this book that the reader review either American cancer mortality data prior to 1950 or incidence data. For the interested reader, however, this section reviews these data. More detailed presentations are made in the references listed at the end of the appendix.

A COMPARISON OF CANCER MORTALITY 1930–50 and 1950–75

The intent of this section is to compare the extent to which 1930–50 and 1950–75 cancer mortality trends are similar. The 1930–50 data is referred to in the following: (1) Gordon, Crittenden, and Haenszel (1961) present and interpret the data; (2) Lilienfeld, Levin, and Kessler (1972) present and review the 1930–71 data along with selected incidence data; and (3) Devesa and Silverman (1978) present and interpret age-adjusted mortality and incidence data for 1935–74. These studies reveal that the trends for 1950–75 are generally similar to 1930–50 trends with some interesting exceptions, notably esophagus, prostate, and bladder cancer and leukemia.

Lung, pancreatic, and brain and other central nervous system cancer mortality rates have consistently increased since 1930 among all four subpopulations. There are differences in *kidney, larynx,* and *female breast* cancer mortality rates during the 1930–75 period among the four populations. Kidney cancer mortality rates have increased more among males than among females; larynx rates have increased more among nonwhites than among whites; and white female breast rates have been relatively stable, while nonwhite female rates moderately increased.

Disaggregated *uterine* cancer mortality data were not available prior to 1950. The total uterus data calculated for 1930–50 exhibit the long-term declining trend characteristic of 1950–75, which, in turn, suggests that factors other than improved screening methods and more frequent hysterectomies account for some of this decrease.

For eight other types of cancer there is a distinct difference between the 1930–50 and 1950–75 mortality trends. These differences are particularly obvious for some of the digestive system and some of the genital organ cancers. Since nonwhite male *esophagus* cancer mortality rates were three times as high as white male rates during 1970–75, it is interesting to note that white male rates were higher than nonwhite male rates until 1945. White male rates increased more than 15 percent during 1930–50 and thereafter have been stable. In comparison, nonwhite male rates increased 500 percent between 1930 and 1950 and about 70 percent between 1950 and 1975.

White male *stomach* cancer mortality rates were also higher than non-white male rates until 1950. Nonwhite male rates exceeded white male rates by 1950 because among the four populations only nonwhite male rates increased between 1930 and 1950.

Rectal and *intestinal* cancer were aggregated to reduce disease classification problems. All four populations exhibited increasing rates between 1930 and 1950. During the subsequent 26 years, nonwhite rates increased, continuing the 1930–50 trend. However, white male and both female rates decreased, reversing the trend and starting a declining trend. Cutler and Devesa (1973) suggest that a more Westernized diet associated with an increasing socioeconomic status of blacks helps account for the continuing increase among blacks.

Two types of genital cancer, *ovary* and *prostate*, manifested interesting differences between 1930–50 and 1950–75. Ovarian cancer mortality rates increased far more before rather than after 1950. The 1950–75 data show an increasing gap between nonwhite and white male prostate cancer mortality. Although white male rates increased more than 10 percent during the 1930's and 1940's, nonwhite male rates doubled. Nevertheless, before 1950, white male rates were higher than nonwhite male rates. Since 1950, white male rates have stabilized, whereas nonwhite male rates have increased more than 40 percent. Ernster, Selvin, and Winkelstein (1978) argue that the gap between nonwhites and whites is a manifestation of specific cohorts, especially the population born during 1896–1900. They suggest that the difference between whites and nonwhites will decrease as these elderly cohorts decrease.

The final two types for which data were available, and which manifest interesting differences between pre- and post-1950 trends, are *bladder cancer* and *leukemia*. During 1950–75, bladder cancer mortality rates either decreased or stabilized among the four subpopulations. In contrast, during 1930–50, nonwhite female rates increased about 40 percent and nonwhite male rates increased more than 30 percent. Changes in leukemia cancer mortality rates since 1950 range from a slight decline for white females, to small increases for white males, to more than 30 percent increases for the two nonwhite populations. During 1930–50, leukemia mortality rates substantially increased for all four populations, compared to the major difference between whites and nonwhites since 1950.

The important changes that have occurred during 1930–75 can be summarized by the *all sites* category. During 1970–75, nonwhite male rates were clearly the highest and were diverging from the other three populations. White male rates were the second highest among the four populations. Male rates were about 50 percent higher than female rates.

Four decades earlier, the order had been almost completely reversed! White females had the highest rates, white males the second highest rates, nonwhite females the third highest rates, and nonwhite males the lowest rates by far. Indeed, the nonwhite male all sites, age-adjusted mortality rate was only about 60 percent of the average of the other three populations in 1930. By 1945, the nonwhite male rate was 75 percent of the average of the other three populations. Not until 1955–59 did nonwhite male rates pass the other three populations. The marked reversal of the nonwhite male and white female order is clearly the single most interesting observation derived from comparing the 1930–50 and 1950–75 trends. A second more general perspective is that the most pronounced changes usually occurred between 1930 and 1950, particularly for the digestive system and genital organs.

A COMPARISON OF INCIDENCE AND MORTALITY TRENDS, 1937–76

Three types of relationships between incidence and mortality trends were observed for the period 1937–71. At one end of the spectrum, cancer of the pancreas, ovary, stomach, and esophagus (white population) manifested very similar incidence and mortality trends. At the other end of the spectrum, male bladder and white population cancer of the larynx, prostate, uterine corpus, and thyroid exhibited marked divergence between incidence and mortality trends. The third and largest group of cancers show incidence and mortality trends that were generally parallel with a tendency toward divergence since the mid-1960's. Due to better survival rates for whites than for nonwhites, greater differences between incidence and mortality trends were observed for the white than for the nonwhite population. There is evidence that some long-standing cancer trends may have changed during the 1970's. The evidence is not strong and will have to be consistently observed into the early 1980's before the trends will be widely accepted.

Four cancer incidence surveys have been made in the United States: 1937–39, 1947–48, 1959–61, and 1973–76 (Dorn and Cutler, 1955; Cutler and Devesa, 1973; Cutler, Scotto, and Devesa, 1974; Cutler and Young, 1975; Devesa and Silverman, 1978; Pollack and Horm, 1980). The surveys were not random and are not completely comparable. The first two included data only for a single year in each geographical area. They were carried out in metropolitan areas because it was assumed that the data would be most reliable in the cities. The third national survey included seven metropolitan areas and two states during 1969–71. The first three surveys have seven metropolitan areas in common and include more than 5 percent of the national population. The first three surveys, thus, have an urban bias and all the ethnic, dietary, other personal, occupational, and local environmental factors associated with an urban bias. In addition, some data were lost, and non-melanoma skin cancer data were not routinely collected. The fourth survey, 1973–76, has the scientific and diagnostic advantages of modern medicine and epidemiology. However, it only has four areas in common with the third survey.

Due to a controversy that has erupted around changes in cancer incidence during the 1970's and substantial differences between the first, second, and fourth survey, the author has compared incidence and mortality trends in two parts: 1937–71 (first three surveys) and 1969–76 (third and fourth surveys).

Incidence and Mortality Trends, 1937–71

There was a close parallel between incidence and mortality trends for the aggregate *all sites* category. Nonwhite males manifested the highest increases in incidence, white males exhibited a much smaller increase, nonwhite female rates were relatively stable, and white female incidence rates declined from the late 1930's to the early 1970's. The nonwhite male, all sites, age-adjusted cancer mortality rate passed the white male rate during the mid-1950's; the nonwhite male, all sites, age-adjusted cancer incidence rate passed the white male rate during the mid-1960's. The pattern of divergence between males and females and between non-white males and the other three populations is apparent in the incidence as well as the mortality data.

A very close parallel was found between cancer mortality and incidence of cancer of the *digestive* system. Lilienfeld, Levin, and Kessler (1972) classify the major types of digestive system cancer as having superimposed trends (female esophagus, pancreas), similar shaped trends, but with incidence rates less than twice mortality rates (stomach), or similar curves with incidence rates two to three times mortality rates (male esophagus, small intestine, rectum, liver).

Stomach cancer incidence, like the mortality rates, consistently declined. A very slight divergence between incidence and mortality can be seen during the 1969–71 survey, *Liver* cancer incidence and mortality rates were also clearly parallel. In comparison to most of the other digestive tract cancers, the combined *intestine* and *rectum* category exhibited a much wider gap between incidence and mortality for all four populations, an observation that is consistent with improved survival rates. White female incidence and mortality rates both decreased during 1937–71. White male incidence increased while mortality decreased. Nonwhite male and nonwhite female incidence and mortality rates both increased.

Less than 5 percent of patients with cancer of the *pancreas* survive three years. As a result, incidence and mortality rates have been virtually superimposed since the mid-1950's. Survival rates for *esophageal* cancer are only slightly better than those for pancreatic cancer. Accordingly, the shapes of the incidence and mortality curves are again virtually identical among whites. Insufficient incidence data precluded a comparison for nonwhites.

The two respiratory tract cancers for which comparative data were available are larynx, and bronchus and lung. Incidence and mortality rates of *larynx* showed increasing differences in every survey. Incidence rates increased, whereas mortality rates were relatively stable. Again, nonwhite comparisons of incidence and mortality rates were not possible due to insufficient incidence data.

Lung cancer incidence rates as well as mortality rates show the highest increases for all populations. Surprisingly, mortality rates have been catching up to incidence rates for all populations since the 1960's. Cutler and Devesa (1973) explain this observation by better reporting of cause of death. Nonwhite mortality and incidence rates passed white male rates during the early 1960's. Cutler and Devesa (1973) hypothesize that the higher increases among nonwhites than whites may be due to metropolitanization of former Southern, rural blacks. Incidence rates were higher among white females than nonwhite females. However, white female lung cancer survival rates were higher than those of black females. Therefore, their mortality rates were about the same.

White and nonwhite *female breast cancer* incidence trends were parallel. Due to higher survival rates for whites, nonwhite female mortality rates increased, white rates stabilized, and the gap between whites and nonwhites almost disappeared by 1971.

Incidence and mortality data were available for three genital organs: prostate, uterus, and ovary. White male *prostate cancer* incidence and mortality trends were noticeably different. Incidence rates increased and mortality rates stabilized due to improved survival rates. Nonwhite male incidence rates and mortality rates both consistently increased, leading to divergence between the two populations.

White and nonwhite combined *uterine cancer* incidence rates and mortality rates decreased. Mortality rates decreased more than incidence rates, a trend that interestingly began even before vaginal cytology screening programs became widespread (Cutler and Devesa, 1973). *Ovarian* cancer incidence and mortality rates have been parallel since the early 1950's.

The urinary tract cancers demonstrated marked divergence between incidence and mortality and between the populations. Male incidence rates increased for *bladder* and *kidney cancer*. In comparison, white male bladder mortality rates stabilized and nonwhite male rates increased. Kidney cancer mortality rates increased, but more so for nonwhite males than for white males. Overall, during 1937–71, the difference between white male and nonwhite male bladder and kidney cancer mortality rates decreased due to greater improvement in survival rates among white males than among nonwhite males.

White female bladder incidence rates have decreased since the late 1940's. Mortality rates have plunged more than incidence rates. Nonwhite female incidence rates have also decreased and mortality rates have slightly declined, although not as much as incidence rates. Improved survival rates are apparent for white females, but not for nonwhite females. Improved survival rates are also apparent for white female kidney cancer. Nonwhite female incidence data were not available.

Incidence data for leukemia, Hodgkin's disease, and non-Hodgkin's lymphoma have been available since the late 1940's. Beginning with *leukemia,* there was no consistent relationship between incidence and mortality prior to the late 1960's. White male and white female mortality rates were climbing faster than incidence rates during the late 1940's and 1950's. Beginning in the mid-1960's, mortality began to decline and diverge from incidence due to improved survival of children with acute lymphatic leukemia. The nonwhite data show rising incidence and mortality rates.

The *Hodgkin's disease* data (for white only) show increasing incidence and markedly decreasing mortality since the late 1960's. The *lymphoma* data exhibit increasing incidence and mortality, a widening gap between them as survival rates increase, and nonwhite rates approaching white rates by the early 1970's.

The final two types of cancer for which incidence data (white only) were available are *melanoma* and *thyroid cancer*. Their trends are markedly different. Melanoma is characterized by rapidly increasing incidence and mortality rates and a small gap between them due to improving survival rates. Thyroid cancer incidence rates have increased. However, mortality rates have continuously and markedly decreased due to improved treatment.

Incidence and Mortality Trends, 1969–76

The above analysis deliberately stopped before the 1973–76 period because a controversy erupted about the most recent incidence data. Pollack and Horm (1980) compared the 1969–71 and the 1973–76 American cancer incidence data. Mortality data were used for comparisons. The 13-page paper contains many caveats about the quality of the data. For example, the authors note that there is a lack of proven comparability between the areas in the two surveys and between the two surveys and the United States. Furthermore, they note that the populations at risk were estimated for most of their study period. They rule out the development of black population incidence trends because of data shortcomings. Pollack and Horm (1980, p. 1100) state that "this analysis has focused more on the methodology of and problems in the analysis of trends over the period of time covered by two nonrandom samples of the U.S. populations than on the trends themselves."

The warnings have been ignored by many because the data standardized to the 1970 U.S. population show that some white male incidence rates increased more rapidly than previously and that white female rates have begun to increase. The trends reported in the 1969–76 study closely parallel the 1935–74 study with some interesting exceptions. Annual

average percentage changes in lung, ovary, and prostate cancer, melanoma, and cervical and stomach cancer seem similar to long-term incidence trends. The exceptions are cancer of the colon, rectum, and uterine
corpus. In these cases a reversal of longstanding trends is indicated. When
considered with continued increases of lung and other increasing types
and reduced rates of decrease of some types, the net result is a 1.3 percent
annual average increase in incidence for white males and 2.0 percent
annual average increase for white females. Increases in mortality during
1969–76 reported by Pollack and Horm for white males and white
females were 0.9 and 0.2 percent, respectively.

Although the trends derived from 1969–76 data have received a good
deal of publicity because of the political implications of finding an upturn
in cancer rates at a time when environmental regulation is being chal-
lenged (Smith, 1980), the Pollack and Horm data yield inconclusive
trends about a brief period. The 1976 rates are higher than 1969 rates,
but incidence and mortality rates have not increased every year since
1969. For example, if 1969 is the starting year, then white female mortality rates have increased. If 1971 is the starting year, then white female
mortality rates have decreased. A similar up and down pattern charac-
terizes the black female mortality data. The up and down pattern is even
more evident among the specific types of cancer and is also apparent in
the incidence data, though far less for the incidence data than for the
mortality data. The combination of a short time period and annual variations is also evident in the wide confidence bands around the average
annual percent changes in mortality and incidence. Forty percent of the
confidence interval bands around the average mortality and incidence
trends for 1969–76 are positive at one limit and negative at the other.

In this author's opinion, the 1950–75 mortality data provide a much
better, albeit more conservative, perspective upon which to determine
how rapidly cancer rates have been changing. Briefly, the data are
scanned for significant changes in rates. If the 1970's has witnessed a
marked change in cancer mortality, then the intervals between statistically significant changes in cancer mortality rates should be different
from those observed during the 1950's and 1960's. To be conservative,
statistical significance is defined as 99 percent confidence that a change
has occurred. As an example, it will be determined if one can be 99 percent confident that a significant change occurred between 1950 and 1951
in white male cancer mortality rates. The 1950 age-adjusted rate for
white males was 158.91/100,000, and the standard error of the rate was
1.002. Multiplying the standard error by 2.58 and adding that rate to

158.91 yields 161.50, the upper limit 99 percent confidence limit (158.91 + 2.59 = 161.50). The lower limit 99 percent confidence limit for 1951 is 157.21 [159.77 − 2.58 (.992) = 157.21]. If the two confidence limits do not overlap, one can assume that a significant change occurred. One cannot be 99 percent confident that the 1951 rate is higher than the 1950 rate. It is not until 1953 that one can be 99 percent confident that a statistically significant increase has occurred (Table A3–1).

If there has been a significant upswing in cancer mortality since 1969, then the intervals between significant increases should be shorter during the 1970's than earlier. With respect to males, there is some evidence to support the upswing conclusion. The most rapid increases in white male cancer mortality were at the beginning (1950–55) and the end (1970–75) of the study period. The most rapid nonwhite male increase was during the late 1960's.

With respect to females, there is weaker evidence of an increase, but definite evidence of stability. The nonwhite female cancer mortality rate

Table A3–1 Statistically Significant Changes in Cancer Mortality Rates 1950–75

99% Confidence Limit of Age-Adjusted Rates/100,000			
Population	Time Period	Initial Rate	Interval (Years)
White Male	1950–52	158.91	3
	1953–55	164.89	3
	1956–62	169.97	7
	1963–66	175.89	4
	1967–70	182.51	4
	1971–73	188.16	3
	1974–75	192.87	2 or more
White Female	1950–53	140.35	4
	1954–57	135.59	4
	1958–63	130.29	6
	1964–75	125.44	12 or more
Nonwhite male	1950–53	145.42	4
	1954–58	163.17	5
	1959–64	179.75	6
	1965–67	196.74	3
	1968–72	213.87	5
	1973–75	231.61	3 or more
Nonwhite female	1950–75	145.57	26 or more

has not significantly (at the 99 percent level) changed from 1950 through 1975. The most pronounced difference is between 1950 (145.57) and 1970 (134.17), a difference significant with 95 percent confidence. Four significant decreases have been observed among white females. Most of these were during the 1950's and 1960's. Since the mid-1960's, the white female rates have stabilized. There has been an increase, albeit inconsistent, in the white female and nonwhite female rates since 1973. However, the change is not strong enough to be statistically significant at the conservative 99 percent confidence level.

AGE-SPECIFIC MORTALITY TRENDS:
LUNG, INTESTINE, AND RECTUM; FEMALE BREAST;
AND BLADDER CANCER—1930–50 AND 1950–70

Four types were chosen to illustrate differences between urban and rural areas. Age-specific mortality rates were presented in Chapter 3 of this volume for the population 35 years and older. This section considers the extent to which the 1930–50 trends are similar to or different from the 1950–75 trends for these four types.

Overall, age-specific mortality trends from 1930–50 are quite similar to 1950–75 trends. The generalization is particularly true of *lung cancer,* for which every age group (35$^+$) has increased in mortality since at least 1930. The most pronounced increases have been in the 55- to 84-year-old populations.

The 1930–50 data extend the 1950–75 trends for the combined *intestinal* and *rectal cancer* category. The decline in white male and white female age-specific rates for the 35- to 64-year-old population began around 1940. The increase of mortality rates among elderly nonwhites (65$^+$) began as early as 1930.

One of the interesting observations about white *female breast cancer* was the decline in rates in the 70-year-old and older age group. The 1930–50 data suggest that this decline began about 1940.

Bladder cancer is the last of the four specific types to be considered. The white male trends for 1950–75 were decreased in the 45- to 64-year-old age groups and increased in the 70 and older age groups. These trends were evident as early as 1930. The 1950–75 age-specific data for nonwhite males is best described as inconsistent. Some of the age-specific rates increased, while others decreased. The 1930–50 data show far greater consistency. All manifested increases except for age group 75–84. The last age group increased during 1930–40, decreased during the 1940's, and increased during the 1950's.

REFERENCES

Cutler, S., and Devesa, S. 1973. "Trends in Cancer Incidence and Mortality in the U.S.A." In R. Doll and I. Vodopija, eds. *Host Environment Interactions in the Etiology of Cancer in Man*, IARC Sci Pub No. 7, Lyon, France: IARC, pp. 15–34.

Cutler, S., Scotto, J., and Devesa, S. 1974. "Third National Cancer Survey—an overview of available information." *J. Natl. Can. Inst.* 53: 1565–1575.

Cutler, S., and Young, J., Jr. 1975. *Third National Cancer Survey: Incidence Data*, NCI monograph 41, Washington, D.C.: U.S. Government Printing Office.

Devesa, S., and Silverman, D. 1978. "Cancer incidence and mortality trends in the United States: 1935–1974." *J. Natl. Can. Inst.* 60(3): 545–571.

Dorn. H. and Cutler, S. 1955. *Morbidity from Cancer in the United States, Part I*, NCI monograph 29, Washington, D.C.: U.S. Government Printing Office.

Ernster, V., Selvin, S., and Winkelstein, W., Jr. 1978. "Cohort mortality for prostatic cancer among United States nonwhites." *Science* 200: 1165–1166.

Gordon, T., Crittenden, M., and Haenszel, W. 1961. "Cancer mortality trends in the United States, 1930–1955." *End Results and Mortality Trends*, National Cancer Institute, cancer monograph 6, Washington, D.C.: U.S. Government Printing Office, pp. 131–298.

Lilienfeld, A., Levin, M., and Kessler, I. 1972. *Cancer in the United States*, Cambridge, Mass.: Harvard University Press.

Pollack, E., and Horm, J. 1980. "Trends in cancer incidence and mortality in the United States, 1969–1976." *J. Natl. Can. Inst.* 64(5): 1091–1103.

Smith, S. 1980. "Government says cancer rate is increasing." *Science* 209: 998–1002.

Author Index

267

Subject Index